CATEGORY 5

CATEGORY 5

THE STORY OF CAMILLE
Lessons Unlearned from
AMERICA'S MOST
VIOLENT HURRICANE

ERNEST ZEBROWSKI
& JUDITH A. HOWARD

The University of Michigan Press   *Ann Arbor*

Copyright © 2005 by Ernest Zebrowski and Judith A. Howard
All rights reserved
Published in the United States of America by
The University of Michigan Press
Manufactured in the United States of America
♾ Printed on acid-free paper

2008   2007   2006   2005   4   3   2   1

A CIP catalog record for this book is available from the British Library.

Library of Congress Cataloging-in-Publication Data

Zebrowski, Ernest.
    Category 5 : the story of Camille, lessons unlearned from America's
most violent hurricane / Ernest Zebrowski and Judith A. Howard.
       p.   cm.
    Includes bibliographical references and index.
    ISBN-13: 978-0-472-11525-9 (cloth : alk. paper)
    ISBN-10: 0-472-11525-1 (cloth : alk. paper)
    1. Hurricane Camille, 1969.   2. Gulf Coast (Miss.)—History—20th century.
3. Hurricanes—North Atlantic Ocean.   4. Cyclones—Tropics.   I. Title:
Category five.   II. Howard, Judith A.   III. Title.

QC945.Z43   2005
363.34'92209762—dc22   2005028583

frontispiece: View of Camille by the ESSA 9 satellite at 2:57 p.m., August 17, 1969.
Camille was the first hurricane to be tracked continuously by satellite from birth to
death. *(USACE, Mobile District.)*

This book is dedicated to the loving memory of
Hazel M. Howard
1923–1992
&
Willie M. Howard
1918–1995
who would understand why it is also dedicated
to Prince
and to the memory of:
Tiny, Cutie, Bridget, Sadie, & Saathi

If I could have one part of the world back the way it used to be, I would not choose Dresden before the firebombing, Rome before Nero, or London before the blitz. I would not resurrect Babylon, Carthage, or San Francisco. . . . I want the Mississippi Gulf Coast back the way it was before Hurricane Camille.

—Elizabeth Spencer, *On the Gulf* (1991)

It's hard to believe the damage could be so great. My honest feelings are that there could never be an accurate record in the history books as to what has gone on here. Pictures will show part of it. Words will tell part of it. But it will never be accurately described.

—Ed Tinsley, Virginia State Police

Preface

The 1960s was a long decade of assassinations, race riots, environmental disasters, and war, all beamed directly into American living rooms. The Cold War loomed in the shadows, threatening at any moment to flare into a nuclear Armageddon. The Supreme Court was transforming race relations and gender roles. Congress passed civil rights legislation over the fierce opposition of many southern senators. An iconoclastic generation of youth shocked its elders. Traditional boundaries and social categories were being upset in every corner of American life.

In the face of these changes, many rural American communities receded into themselves. Indeed, some of these regions, particularly in the South, had never been particularly well-connected to the outside world in the first place. Many preached a kind of home-grown isolationism, often expressed in the language of American self-reliance and a wary opposition to centralized government.

Then, in August 1969, Mother Nature yanked three such regions—Plaquemines Parish, Louisiana; the Mississippi Gulf Coast; and Nelson County, Virginia—sharply out of their isolation by delivering the most powerful hurricane to strike the American mainland in recorded history. Camille, as the storm was called, had the highest sustained winds—sea-level measurements topped out at 172 miles per hour, but winds as high as 201 miles per hour were extrapolated from aircraft data—of any hurricane ever to strike the United States. Her storm surge—officially measured at 24.6 feet inside a surviving structure, but almost certainly reaching more than 28 feet—also set a record, as did the thirty-two inches of rain she dumped on rural Virginia in just six hours. The Virginia deluge triggered some 150 landslides and generated floods of almost Biblical proportions. The memory of these events was so terrible that, almost forty years later, many long-term residents of the stricken regions still marked time in terms of "before Camille" and "after Camille."

It was almost much worse. Camille turned north in time to narrowly miss New Orleans. Even so, hundreds of people were killed, and thousands more lost everything they had. Recovery of the stricken communities took decades. In the summer of 2005, bare concrete foundation slabs—vestiges of Camille—still dotted many Gulf Coast neighborhoods.

While geographic, sociopolitical, and cultural factors affect the impact of every disaster, such considerations took on even more significance than usual in the aftermath of Camille. As the enormity of the catastrophe sank in, local officials were forced to temper their mistrust of outsiders and accept at least some offers of federal assistance. This change in attitudes was not universal, nor was it equally effective across the board. Even as Camille laid bare the myth of rural self-reliance and opened these once-isolated regions to outside help, the old culture of individualism still claimed justification in the inevitable glitches of the recovery effort.

Although individual initiatives indeed proved necessary, they were not a sufficient response to the overwhelming challenges. It takes more than a cluster of individuals, more even than a local community, to successfully respond to nature on a rampage. And so Camille secured, for the next three decades, Washington's role in providing disaster mitigation and relief in an increasingly complex society. The result was a raft of new federal programs, improvements in emergency management planning (eventually leading to the creation of FEMA), increased support for scientific and engineering research (resulting in new building codes and the adoption of the Saffir-Simpson potential damage scale, among other innovations), and a clearer understanding of the psychosocial consequences of disaster trauma.

Then, just as we were completing this book in August 2005, nearly 90,000 square miles of the central Gulf coast were devastated by another monster hurricane. Katrina, an enormous Category 4 storm, was immediately dubbed by some old-timers as "Camille's evil daughter."

Katrina was in many respects a replay of Camille, which had made essentially the same landfall back in 1969. And while the historic catastrophe of August 1969 had supposedly awakened everyone—Congress, meteorologists, civil engineers, relief workers, and emergency response specialists—to the multifaceted complexities of disaster management, many of the lessons learned from that experience about the hard work of disaster preparedness, evacuation, and emergency response seemed to have been forgotten.

In response to the events of August and September 2005, we have revised the concluding chapter of this book to speak to a few of the parallels between the two hurricanes. But we have purposely limited our analysis of the response to Katrina, on the assumption that it will take several years to sort out responsibility for the tragic blunders of that effort. The main focus in the pages that follow is Camille, and the many lessons learned—some, sadly, later unlearned—from that terrible storm. We have decided to let the story of Camille speak for itself, essentially as we wrote it prior to the ravages of Katrina. We leave it to you, the informed reader, to draw your own conclusions.

Acknowledgments

We are grateful to the many dozens of people—community leaders, librarians, professional colleagues, students, military men, and everyday citizens in the communities we visited—who generously took the time to talk with us and to point us in the direction of various resources. We are especially indebted to those survivors of Camille and their family members who assisted us, and for opening their hearts to us by sharing memories that were often still painful, sometimes to the point of tears, even all these years later. It is impossible to fully depict what it must have been like to survive the chaotic fury of Camille; while we have strived for accuracy, we apologize if we have introduced any misconceptions or errors of fact into their stories, and offer them our sincerest apologies.

To include all of the personal accounts we gathered, unfortunately, would have swelled this book to the size of a small encyclopedia. In fact, we ultimately cut more than 100 pages out of our initial manuscript to arrive at this book. This should not be construed as disregard for the contributions of those whose stories had to be left out; every one of them has contributed in some manner to what follows, although they do not always appear by name.

We are deeply indebted to Linda Frazier, Lina Iu, David Lopez, Linn Joslyn, and Cleopatra Mathis for reading early drafts of this material and for offering valuable suggestions and to Sandy Mansfield for her unwavering support and encouragement. We greatly appreciate the understanding of our friends and family who too often heard, "I can't. I'm working on the book." Last, but hardly least, we are grateful to our insightful and dedicated editor, Jim Reische, for his enthusiastic support of this project. His good humor saved us on more than one occasion.

Ernest Zebrowski and Judith A. Howard
September 18, 2005

Contents

Illustrations following page 116

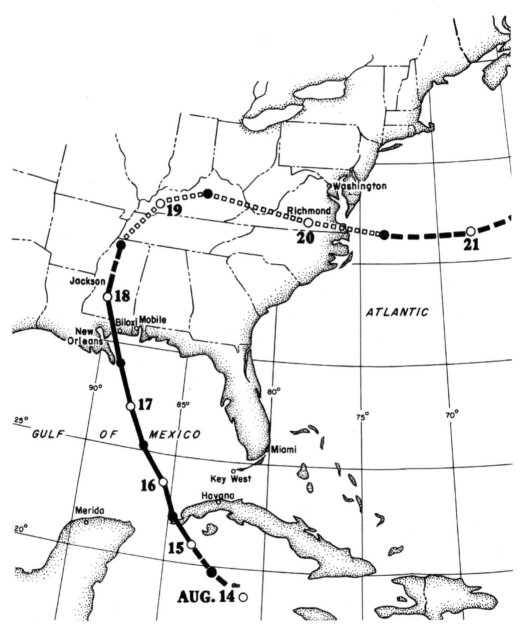

Camille's path inland. *(Adapted from USACE, Norfolk District.)*

CHAPTER 1

GRIM NEWS

Monday, August 18, 1969

Josephine Duckworth paced in the living room of her upscale home
in Jackson, Mississippi. The yard was littered with tree limbs, mangled porch furniture, and other debris that had originated who knew
where, but it wasn't that mess that distressed her. Her frantic
thoughts were on Ben, her twenty-three-year-old son. She hadn't
heard from him since six o'clock the previous evening.

Born in Alabama, Josephine fit every positive stereotype anyone
might have of a classic southern belle. Blonde, petite, and ever the
lady, invariably dressed as if expecting to be photographed for posterity, she was loved by all who knew her. No new acquaintance could
fail to be captivated by her charm, her magnetic smile, and her lilting
southern accent. Ben had often joked, both to his friends and teas-
ingly to his mother directly, that she was the only woman in the
world who could turn her birthplace, Troy, into a three-syllable
word.

She'd spent a sleepless Sunday night pacing, praying, and dialing
one phone number after another. She'd left messages for the governor and the adjutant general that her son and two dozen others had
stayed at the Richelieu Apartments in Pass Christian. Now, Monday
morning, the newscasts still reported nothing about Camille's impact
on the coast. In fact, the Mississippi Gulf Coast, 140 miles to the
south, had been cut off from all communication.

Her husband, Hubert, shut the front door behind him and peeled
off his dripping raincoat. The storm had tapered to a drizzle, and he'd
just confirmed that there was no serious damage in their own neigh-

borhood. He needed to get to his office. There would surely be a lot to do after this one.

A vice president of Borden Inc., Hubert Duckworth was responsible for all of the company's operations in the state of Mississippi. And, indeed, he would soon learn that the hurricane had created a whole host of problems for his managers and staff—including damage to production facilities, statewide complications in the distribution of Borden's dairy products, and devastating losses of dairy herds in the southern counties. In fact, bad news about additional problems would continue to arrive for weeks to come. Josephine, however, was not interested in Borden at the moment. She reminded Hubert, with no mistake in her voice, that while he'd been sleeping through the night she had been wide awake and doing all of the worrying for the both of them. She was not about to let him leave the house until they heard from Ben.

Although Hubert was not one to show much emotion (in Josephine's words, he wasn't a "folksy" kind of fellow), he tried his best to console her. They had raised a very smart son, he assured her. Certainly Ben was okay. The problem, he explained, was simply that hurricanes blow down trees and that phone lines are no match for falling trees.

Josephine tried the phone again only to get another busy signal. Not a single telephone in Harrison or Hancock County remained active, and the trunk lines were clogged with thousands of futile attempts to get through. She turned up the volume on the television, but the local newscasters still babbled about the minor damage in Jackson while only alluding to unconfirmed reports of "major damage" on the coast. In her mind, Josephine rolled over and again the reasons why Ben had stayed behind in the face of the mandatory evacuation order. Not just the reasons he'd told her on the phone yesterday evening but the other possible reasons as well.

Hubert, meanwhile, dialed Robert Pendleton, a private investigator who had done various jobs for Borden over the years. When that brief conversation ended, Hubert assured Josephine that they'd know something in a couple of hours. Pendleton's New Orleans office had already chartered a helicopter, and some of his boys were fixin' to take off for an aerial inspection. Pendleton would tell them to check out the condition of the Richelieu, and as soon as they did, he'd phone the Duckworths with the news.

The call came around noon. Hubert snatched up the receiver, listened, then broke into a grin. Josephine sank into the sofa and sighed in relief, clutching her chest. The flyboys had radioed their report, and the Richelieu Apartments had stood up just fine. Although there was terrible devastation nearby, Ben was surely safe. The Duckworths hugged, then went to the kitchen to stir up brunch while waiting to hear from Ben directly. Hubert began thinking about how he would prioritize his responsibilities at the office in the aftermath of the disaster.

The report from the helicopter, however, turned out to be a cruel mistake. So many landmarks in Pass Christian had been flattened that the pilot and observers had mistaken another structure farther inland—it was never clear which one—for the Richelieu Apartments building. In fact, so little physical evidence of the Richelieu had survived that it was impossible to pinpoint the former site of that place from the air. Later that afternoon, Pendleton phoned Hubert back with the bad news. He offered the Duckworths his prayers and told them that there might be something on the *CBS Evening News.*

This was a time, mid-1969, of widespread distrust of public officials. The peace and civil rights activists had split the nation, the women's movement was gaining steam, and President Nixon's inconsistent public statements about the war in Vietnam had disillusioned so many that a widely distributed poster posed the question in banner lettering under his photograph, "Would you buy a used car from this man?" In that atmosphere of discontent, Walter Cronkite, the fatherly anchor of the *CBS Evening News,* held his team of reporters and writers to impeccable standards of journalistic veracity. A public opinion poll actually identified Cronkite as "the most trusted man in America."

That evening, August 18, 1969, Cronkite stared somberly into the camera and told the nation of the terrible destruction of the Mississippi Gulf Coast by a hurricane named Camille. The National Hurricane Center (NHC) had reported winds of 172 miles per hour that gusted to over 200, making Camille the most intense hurricane to strike the U.S. mainland in the nation's history. Human tragedy bordered on the unspeakable. As the screen switched to aerial footage of the devastation, Cronkite commented, "This is the former site of the Richelieu Apartments, where twenty-three people laughed in the face of death, and twenty-three people died."

The Duckworths were devastated.

Cronkite went on, but they heard little more. As Hubert tried to be stoic in accepting the news, Josephine exploded in tears. She bargained with God, praying that this latest information was just another mistake. She vowed that she would not abandon hope until they had indisputable firsthand confirmation of Ben's death. After all, they'd lived on the coast near Biloxi until Ben was in eleventh grade, so the boy certainly knew about tropical storms and how serious they could be. Surely, of all people, their son wouldn't have been foolish enough to attend a "hurricane party"! But through the rest of the evening and through the night, no calls came with any further news.

The following morning, as always, Hubert donned a white shirt and a tie. His first stop was the bank, where he withdrew five hundred dollars. The local funeral director had advised him to carry the cash on his trip to Pass Christian, because it might be needed to assure that Ben's body was transported to Jackson quickly and with due care. They would have an open-casket viewing if possible, but that would depend on the condition of Ben's corpse when it arrived. Josephine, meanwhile, remained at home, waiting and praying for the phone call from Ben that, in her mind, could possibly come at any minute.

After picking up one of Ben's friends, Charles Edward Barranco, Hubert drove to the home of his daughter, Marian. His son-in-law, Bill—the Reverend Dr. Bill Duncan, a Baptist minister—was waiting at the driveway. He asked Hubert to open his trunk so they could load a few things: a couple of bow saws, an axe, a coil of rope, a come-along, a box of miscellaneous tools, work gloves, flashlights, jugs of water, a full gas can, and blankets. Marian scampered up with a bag of sandwiches and, teary eyed, kissed them. Hubert hadn't eaten a sandwich in years—he just wasn't a sandwich kind of person—but he thanked his daughter for her thoughtfulness. It hadn't occurred to him that food and water might be a problem or that they might need to cut their way through obstructions to get to the site of Ben's apartment.

They headed south on two-lane Highway 49. The road had already been cleared of debris, and the trio made the first eighty miles without incident. Approaching Hattiesburg, however, they began to see evidence of wind damage. Not just the fallen branches and scattered trash they had in Jackson but garage roofs lifted, trees uprooted, cars crushed, and farm outbuildings flattened. South of Hattiesburg, with another seventy miles to go, they were stopped at a roadblock manned by National Guard troops. Just a short time earlier, at 11:37 that morning, the governor had declared martial law for the coastal counties.

A young corporal with a rifle slung over his shoulder asked Hubert for his destination. "Pass Christian," he answered.

The young man shook his head. "You can't go there, sir. The hurricane wiped out the roads."

"Did it wipe out 49?"

"No, sir, but it sure did a big-time job on Highway 90. Our orders are to not let anyone through to the coast. What's your business?"

Hubert took a deep breath. "My business, young man, is that my son was just killed there. We're going there to get his body."

The corporal's face ran pale. "I'm sorry, sir . . . ," he stammered. "I can't guarantee that you'll get through. But if you want to try, I sure won't be the one to stop you." He waved them through the roadblock.

The opportunity to speed down an open road was short. After a few miles, they came up behind an incongruously fully loaded Budweiser beer truck. They followed it almost to Gulfport, where it turned east toward Biloxi. Although Hubert, Bill, and Charles Edward didn't know it then, all of those cases of cans labeled as brew were actually filled with water destined for survivors and relief workers. Hubert turned right onto Pass Road. This was the back way into Pass Christian, and he knew it well.

The devastation there was phenomenal, with not a single structure undamaged. They were stopped every couple of blocks by deputies or National Guard troops. The din of chainsaws muddled all attempts at communicating, and Hubert explained his mission over and again. With the few remaining blocks impassable to vehicles, they got out and walked. Despite the August heat, Hubert still wore his tie, his collar still buttoned. It never occurred to him to do otherwise.

Ahead and to the south lay a wasteland, with not a wall left standing within three blocks of the Gulf. They climbed over and through the rubble to the site of the Richelieu, which they could identify only by its swimming pool. Truncated pipes jutted eerily from the foundation slabs; Camille had even claimed the toilets and bathtubs those pipes had supplied. Hubert wandered through the devastation in a semi-stupor, remembering the smiling photographs of Ben, Josephine, and their young granddaughter snapped at this very spot last Christmas. Now, it seemed as if those memories had been part of another world. He prodded himself into focusing on his current mission: he needed to find the morgue and have Ben's body shipped home. Considering the number of lives the storm had claimed in this little town, surely the local authorities had set up a temporary morgue somewhere close.

Hubert noticed a young man in olive drab tramping around nearby, also shaking his head in solemn disbelief. Hubert squinted; the fellow looked familiar. In fact, he was an acquaintance of Ben's, Mike Gannon. It didn't immediately register with Hubert that this fellow had also been living at the Richelieu. He asked Mike if he knew where Ben might have been taken.

"Yes, Mr. Duckworth," Mike replied. "I saw them carry Ben into the high school yesterday morning along with a lot of others. Do you know where that is?"

Hubert nodded and swallowed as he straightened his tie. Southern gentlemen don't cry—at least not in public.

Thirty-three-year-old Mary Ann Gerlach gained near-instant national fame with her remarkable survival story, which she enhanced with additional details each time she was interviewed by yet another reporter. She and her sixth husband, Frederick (or "Fritz," as most people knew him), had lived at the Richelieu. Both had worked night shifts the evening before the hurricane, she as a cocktail waitress and he as a Seabee in the navy. Mary Ann told reporters:

> The first thing that popped in my mind was party time! We all
> got together and decided we were going to have a hurricane
> party on the third floor. I went out and got all kinds of stuff to
> fix, you know, sandwiches and hors d'oeuvres and got a bunch

of stuff to drink. Well, all the Civil Defense people had come up trying to get us out, and the manager and his wife kept telling us, "No need to go, it's ridiculous, just stay here."

Mary Ann and Fritz never did join the group on the third floor. They decided to nap first and were awakened around 10:00 p.m. by thumping sounds from below. The electricity was out by then, and they ventured into the living room by flashlight. To their horror, the Gulf of Mexico was one-third of the way up their second-story picture window, some twenty feet above normal sea level. As they dashed back to their bedroom, the front window imploded, the sea rushed in, and the building shuddered. Years earlier as a new enlistee, Fritz had talked a buddy into passing his swimming test for him, and now that ruse came back to haunt him; he couldn't swim a lick. With waist-deep water swirling around them and their furniture floating, Mary Ann blew up an air mattress she kept for the swimming pool and gave it to Fritz.

Moments later, the rear window shattered and she swam out with the current—smack into a maze of electrical wires. The sea, surging in through the front and out the rear of the apartment, swept Fritz out behind her. She disentangled herself and pushed off from the doomed building. "My legs," she explained, "were real strong, you know, from doing cocktail waitress work for so long." Meanwhile, Fritz drowned. Several days later, his body was found tangled in a tree several miles inland, mangled and decomposed.

Some six hours later, unable to walk and wearing only tattered shorts and the ragged remnants of a short-sleeved sweatshirt, Mary Ann sat shivering in the mud into the morning. She spied a man tramping through the debris and called to him for help. He asked if she had seen his wife. "No, I haven't seen anyone alive but you," she replied. The man stumbled away in a trance, repeating his wife's name over and over.

She huddled, still shivering, for more than an hour before the next person came along, a young man she recognized as a local post office clerk. She shouted to him. The postal clerk and two other men carried Mary Ann to the white high school, where the shop area was being converted to a temporary morgue and most of the rest of the building was sheltering survivors. A few hours later, several National

Guard troops in a LARC transferred her to the Miramar Nursing Home. There, as her wounds were being tended, Mary Ann explained to the nurses that she was the sole survivor of the Richelieu Apartments.

The word quickly got out to the reporters, and as journalists swarmed in to interview her over and again, Mary Ann got better and better at remembering various details of her extraordinary survival story. Nationwide, hundreds of broadcasts and newspapers reported that Mary Ann Gerlach had been the sole survivor out of two dozen revelers at a "hurricane party" in the ill-fated apartment building.

In truth, at least eight others had survived the destruction of the Richelieu. Several of them had heroic motives for remaining there that terrible night, and all suffered consequences as agonizing as Mary Ann's harrowing experience. Camille's "hurricane party," however, has become embedded in American folklore, and perhaps some good has come from that. Wittingly or not, Mary Ann Gerlach raised the consciousness of millions of Americans that hurricanes are not auspicious occasions for partying.

Louisiana, Daybreak, August 18, 1969

New Orleans had been lucky. The local damage reported by the *Times-Picayune* amounted to nothing more sensational than awnings and signs down, power outages in the suburbs, a light plane flipped over at the airport, and minor flooding in the lowest-lying neighborhoods. The emerging news on television over the next few days was the incredible devastation along the Mississippi Gulf Coast, running from Waveland through Bay St. Louis, Pass Christian, Gulfport, and Biloxi, and extending at least as far east as Ocean Springs. It was this story about the disaster just to the east, supplemented with dozens of photographs of the mayhem there, that dominated the news in the Big Easy.

It was several days before any news arrived from Plaquemines Parish, home to twenty thousand people. There was but a single two-lane road linking that hundred-mile-long birdsfoot delta with New Orleans, and neither news reporters nor the evacuated residents were permitted back in the parish beyond the National Guard roadblock at

Myrtle Grove. To the south, the road was obstructed by everything from uprooted trees to whole mangled houses to wrecked drilling equipment and even a 120-foot river barge that had washed over the levee. Below the town of Port Sulphur, where there had recently been a thoroughfare with side streets and businesses and residences, there now stretched a twenty-mile-long lagoon trapped between the river's west levee and the parish's back levee. Hundreds of animal carcasses floated in that grizzly pool, along with the wreckages of homes now teeming with snakes and fire ants. The air buzzed with hordes of hungry mosquitoes.

Although the governor and others inspected the damage from the air, the only news to report about lower Plaquemines Parish was that there was virtually nothing left standing there. And that, essentially, was the substance of the first brief articles in the *Times-Picayune*. With nothing left and nobody allowed in, what more could be written about Plaquemines Parish?

Nelson County, Virginia, August 19, 1969

No hurricane can sustain extreme winds very long after it makes landfall, and by midday on Monday, August 18, Camille's jurisdiction passed from the NHC to the parent National Weather Service in Washington. Now diminished to a rain squall, the storm's remnants swept across western Tennessee Monday afternoon, angled across Kentucky and into West Virginia on Tuesday, then took a southerly turn toward the Blue Ridge Mountains of central Virginia. The rains were not particularly heavy, and the Weather Service predicted that the storm would fully dissipate west of the mountains.

Nelson County sprawls over 471 square miles bordered on the northwest by the four-thousand-feet-high Blue Ridge Mountains and on the southeast by the James River. Three shallow rivers—the Tye, the Piney, and the Rockfish—gurgle across the county on their way to the James. Along the way, they are fed by numerous creeks—Davis Creek, Possum Trot Creek, Hat Creek, Little Joe Creek, and so on— which contribute to the beauty of the landscape even though they are normally brooks just a few feet wide and mere inches deep. The four main roads—U.S. 29 and State Routes 56, 151, and 6—all cross rivers and creeks, and they run parallel to one stream or another for miles at a time. Most of the numerous secondary county roads also follow or cross the streams.

In 1969, Nelson County had a population of about 12,000 and no incorporated towns; Lovingston, the county seat, was home to just 670 residents. The county's single desegregated middle school had 324 students, and its two high schools (one black, one white) enrolled 676. Half of the region was forested, and most of the remainder was farmland or pasture. The only two factories stood near the county lines: American Cyanamide's dye extraction plant at Piney River in the southwest and a soapstone operation at Schuyler in the northeast. All in all, it was a quiet place with virtually no crime—not even a graffiti artist and very few litterbugs. Big news was when the county's single traffic light—a blinking one at that—malfunctioned.

Nelson County never produced a governor, a movie star, or a sports celebrity; the most famous native is Earl Hamner Jr., whose published tales of his boyhood in Schuyler became the basis of the popular 1970s television series *The Waltons*. Of their native son Hamner, the county folks were, and still are, intensely proud. Not so much because of his personal success as a writer but because he had so effectively portrayed their cooperative rural value system.

Although Camille no longer packed hurricane-force winds as she approached this tranquil place, she still toted most of the 108 billion tons of tropical moisture she had vacuumed from the Gulf. Nobody in the National Weather Service had thought much about where all that water would end up or what it might do to those it fell on. Nor did anyone consider what might happen when the remnants of Camille collided with another storm system—a cold front—that was then sweeping into Virginia from the north. Given that separately each system was likely to produce rain, the situation was seemingly as straightforward as it could be: forecast rain for central Virginia. And so the weather report in the Charlottesville newspaper tersely predicted on August 19, 1969, "Rain tonight, clearing tomorrow."

Warren Raines and his brother Carl Jr., ages fourteen and sixteen, spent that summer afternoon bicycling around Massies Mill, a village of forty homes in the upper valley of the Tye River. The stream, whose source lay in the Blue Ridge Mountains a few dozen miles away, was barely shin-deep this time of the year. Against the backdrop of the gray-blue mountain range rolled a vibrant carpet of green meadows punctuated with stands of forest and, lower, a patchwork of bucolic bottomland pastures and orchards, split-rail fences, white farmhouses, and weathered barns with red tin roofs. In every direction, the place looked like a picture postcard.

As the sky clouded over, the brothers stopped by the orchard supply store managed by their father. The rural mailman, Tinker Bryant, pulled up and waved a cheerful hello; he attended the same church as the Raines family, and the boys knew his three daughters. He was singing a song that was high on the charts that summer: "God didn't make them little green apples and it don't rain in Indianapolis in the summertime."

Carl Raines greeted his sons and reminded them he'd be a few hours late getting home. Today was the primary runoff election for the Democratic gubernatorial candidate, and he had volunteered to work at the polls for a while. A Democratic primary was a particularly important event here, given that there was only one registered Republican in all of Nelson County.

After a late supper that evening, Warren, Carl Jr., their brother Sandy, age nine, and their sisters Ginger, seven, and Johanna, eighteen, sat on the front porch with their parents as they often did on summer evenings. Missing was their oldest sister, who was away in Lynchburg. With the election still on his mind, Carl Sr. described to his children how Virginia had an excellent record of honest and effective government—surely the best in the whole nation. Even in hard-fought elections, the politics here were honest. A gentle rain had just begun.

The conversation turned to the newly completed section of four-lane road on Route 29, which bypassed the steep and narrow streets of Lovingston and was surely going to make it easier to get around. Warren brought up something more exciting: the hurricane that had torn up parts of Louisiana and Mississippi two days ago. There had been footage of the destruction on the national news that evening. Everyone on the porch concurred that they couldn't imagine living in a place vulnerable to such devastation.

It was around eleven o'clock, with the rain getting heavier, when the Raines family went to bed.

CHAPTER 2

OF LOVE AND LIFE

No disaster—indeed, no human event—is ever written on a blank slate. Collective knowledge, the local culture, and the consequences of prior social decisions and indecisions all combine to affect the human toll when nature goes on a rampage. When their plans and expectations go awry, leaders, followers, those who would prefer to be left alone, and outsiders—including scientists and government officials—find themselves interacting in new and unfamiliar ways. How effectively they cooperate depends on what they know, what they don't know, and what they may erroneously think they know.

The story of Hurricane Camille began long before that particular storm made its debut in August 1969. In fact, it may be argued that the tale began more than three centuries earlier, when European immigrants—unlike the indigenous (and wiser?) American Indians—started building settlements on storm-vulnerable coasts. Perhaps it was natural for the European colonists to exercise a man-over-nature bravado in their patterns of settlement; after all, it was through that same kind of audacity that their culture had managed to develop the maritime technology that brought them to the New World in the first place. But there was also the matter that, in an era of slow and limited communications compounded by language differences (French, English, and Spanish), harsh meteorological lessons learned by one pocket of coastal dwellers did little to inform people elsewhere who were similarly at risk. This continued to be the case through the nineteenth century and even well into the twentieth.

In 1893, for instance, a hurricane struck Cheniere Caminada, a barrier peninsula near Grand Isle, Louisiana, destroying a palatial resort and the nearby town and claiming as many as twenty-three hundred lives. The cruel lesson of that site's vulnerability was heeded by New Orleans investors and vacationers, and the resort was never rebuilt. A mere seven years later and three hundred miles to the west, however, a similar hurricane flooded most of the geographically comparable city of Galveston, Texas, killing about eight thousand. What effect did the Cheniere Caminada disaster have on mitigating the great Galveston catastrophe of 1900? Apparently none whatsoever.

As communications technology advanced during the twentieth century, so did the timeliness and accuracy of the news about hurricanes. Seldom, however, did such information about disastrous tropical storms mobilize any community to plan for them. True, Galveston reacted to its own catastrophe by building a concrete seawall and bringing in fill dirt to raise the entire city. But the Galveston disaster wasn't *news* to the residents of that municipality; it was part of their direct *experience*.

News is what happens to someone else. And for many decades, that continued to be the general attitude in virtually all of the nation's coastal communities—at least those that hadn't already experienced a recent disaster of their own.

It was in 1957 that Louisianans harboring this attitude of collective aloofness were kicked in the pants. The educational stimulus was a hurricane called Audrey. If there was any meaning to be extracted from the misery of Audrey's victims, it was in everyone's hope that at least some public officials and scientists may have learned important lessons from the calamity. Audrey was a disaster that should not be allowed to repeat itself.

And, indeed, it would be largely because of Audrey that, twelve years later, Camille didn't claim many more lives than she did in Louisiana.

Southwest Louisiana, 1957

On the afternoon of June 24, 1957, which is early in the season for a major hurricane, the New Orleans office of the Weather Bureau picked up a radio message from a Mexican shrimp boat reporting heavy seas in the Bay of Campeche, with winds of forty-five miles per hour and gusts above sixty. That evening, after corroborating reports arrived from several other ships in the area, the Weather Bureau issued its first bulletin that a tropical storm had developed in the Gulf of Mexico about five hundred miles south of the Louisiana-Texas border. The next morning, June 25, with the storm escalating to one hundred miles per hour and inching twenty miles closer, the Weather Bureau posted a hurricane watch for the Texas and Louisiana coasts. The name of that slow-moving storm, the first of the 1957 season, was Audrey.

At 10:00 a.m. on June 26, with the storm 317 miles from the coast and tracking north at 7.8 miles per hour, the Weather Bureau upgraded the hurricane watch to an official hurricane warning, predicting tides of five to eight feet above normal and advising residents to move from low or exposed places. Landfall, the meteorologists figured, would be in about thirty-six hours—late on the evening of June 27—and the most likely spot was Port Arthur on the Texas-Louisiana border. That report was relayed to radio and television stations in southern Louisiana and along the Texas coast as far as Corpus Christi. Meanwhile, residents of southwestern Louisiana noticed a strange phenomenon: thousands of critters making a mass exodus from the lower areas. Some enterprising Cajuns scooped up crawfish by the bucket and stuffed them in freezers, figuring to sell them after the storm. And at Holly Beach, large schools of mullet thrashed in the surf, vainly attempting to swim north.

Early on the evening of Wednesday, June 26, Audrey intensified. She would soon enter the range of Houston's ground-based radar, and despite the primitive state of that technology in 1957, it would be possible to begin tracking the storm from land. Louisiana had no weather radar equipment whatsoever at that time, so the New Orleans branch of the Weather Bureau, which officially "owned" the storm, had to depend on Houston's interpretations of their radar blips. A few hours later, Houston reported that the storm's movement had increased to about eleven miles per hour, which meant that it would strike the coast sooner than previously expected. In its 10:00 p.m. advisory, the New Orleans weather station increased its estimate of tidal flooding on the state's western coast to nine feet but made no mention of a revised time of landfall. Such information would probably have made little difference anyway, given the time of night that advisory reached the radio stations.

Although a few folks from southern Cameron and Vermilion

Parishes had evacuated, most had not. Many hadn't even heard about the approaching hurricane until they got home from work on the afternoon of June 26. As night fell, Audrey was still almost two hundred miles offshore and the skies were clear with no more than a gentle breeze. And given that almost nobody lived in the vast marshlands within a couple of miles of the shore, even those who heard about the newly predicted nine-foot tidal surge figured it couldn't possibly reach many homes. The wind speed was people's only concern; they knew that they might see some wind damage in the next day or two.

Then, around midnight, the New Orleans office of the Weather Bureau got two pieces of horrifying information. First came a report that Audrey had intensified dramatically: she now had sustained winds of 135 miles per hour. Second, she was now racing toward the coast at 16 miles per hour and possibly even faster. Audrey would strike southwestern Louisiana early in the morning and with devastating winds.

Oblivious to the fact that there were but two routes north in all of Cameron Parish and just one way out of southwestern Vermilion Parish, the Weather Bureau issued a warning advising everyone in the low-lying areas to head inland at *daybreak*. By then, it was after midnight and most folks were asleep, confident that the hurricane was still a day away. Nor did it help that one radio station had a shift change that led the incoming announcer to repeat an earlier advisory when the new message slipped to the floor and lay there unnoticed. Many late listeners got the impression that nothing at all had changed in the previous six hours and that the storm had stalled out in the Gulf. Others—particularly those Cajuns who had never developed a strong ear for the English language—had difficulty understanding the broadcasts through the rising levels of static on the AM-band radios that were still the norm in 1957.

But Audrey hadn't stalled. Instead, she had escalated both in wind speed and in forward movement, and she made landfall at the town of Cameron (pronounced "Camrun") around 6:00 a.m., with sustained winds of 145 miles per hour punctuated by higher-speed gusts. The Weather Bureau had also been wrong in its forecast of a nine-foot tidal surge; it was actually twelve to thirteen feet in Cameron. On top of this great bulge of seawater, the wind whipped up waves to a height of another six feet. More than one-third of that parish seat's eleven hundred residents perished in the flooding. Many of the survivors owed their lives to the structural integrity of the courthouse; old photographs of Audrey's aftermath show that building standing battered and windowless, surrounded by a domain of total destruction.

Of Louisiana's sixty-four parishes, Cameron Parish is the largest in land area yet the third smallest in population. Today, some 9,991 people—including 388 blacks, 44 Asians, and 37 American Indians—live on the parish's 1,313 square miles of dry land (another 619 square miles are permanently underwater). At the time of Audrey, the parish was even more sparsely occupied, with a total population of just 6,900, or an average of just five folks per square mile. As a point of reference, through the late 1800s, the federal government defined as "frontier" any region that averaged fewer than six residents per square mile.

To enter the parish from the east or west, one still has only a single choice: the two-lane State Route 82, which, several miles inland, parallels the shoreline for about one hundred miles. Within this stretch, just two roads branch off to the north. If Route 82 floods, many folks are certain to be trapped.

In good weather, it's a picturesque drive. Expanses of high grass dance in anthropomorphic choruses, egrets wade in the swamps, hawks circle above. Here and there stands a cabin surrounded by shallow water, planks spanning concrete blocks for foot access. The high and incongruously modern bridge over the intracoastal waterway offers an unforgettable vista of mile after mile of swamps and bayous. It's hard to envision an angry storm-swept sea surging in so far and inundating not just this road but a huge chunk of countryside extending as far as twenty miles farther inland and more than seventy-five miles east and west. Clearly, this is not a place where folks can afford to dally on news of an oncoming hurricane. Nor is there any question that the lives of everyone here depend on the accuracy and timeliness of tropical storm warnings.

Along this stretch are several *chenieres,* French for "places of oaks." These are long narrow stretches of marginally high ground—at least a few feet above the surrounding marsh—that run parallel to the coast. Chenieres are remnants of beaches that were stranded inland in eons past as the Mississippi River shifted its course west and then back east again. Their most distinguishing feature is their forests of stately live oaks, the crooked spreading branches and dangling Spanish moss forming a magnificent canopy.

Pecan Island, which is actually a cheniere rather than an island, lies about seventy-five miles east of the site of Audrey's landfall. Pickup trucks here outnumber cars, and watercraft outnumber pickups—some backyards stowing as many as five trailered fishing boats. The name "Broussard" is on at least half the mailboxes.

Beneath a magnificent oak in Stephen and Florence Broussard's front yard sprawls a shrine to the Virgin Mary. They are in their mid-eighties now, Stephen tethered to a bottle of oxygen. In 1957, the couple had eight living children, another having died as a toddler. Stephen was considerably more robust then—a husky, hard-working, hard-drinking Cajun who, by all accounts, dispatched his consider-able family responsibilities with vigor.

When asked about Audrey, Stephen's voice cracks. He glances to Florence. Only when she nods does he continue. "We went through twenty-four hours of hell," he says, "and it didn't have to be that way."

At that time, they lived on this same property in a single-story four-bedroom frame house. One son, Brent, was off visiting an aunt far-ther inland, but the other seven children were at home. Stephen stayed up listening to the radio for updates about the hurricane. On the news that the storm was gaining strength, he walked outside and found the wind blowing from the southeast and bending the tree-tops, not a good sign. He went to the bedroom and woke Florence.

Florence, however, was reluctant to rouse the children and crowd them into the station wagon in the wee hours of the morning. Her father would be arriving in a few hours. Why not wait and include him in the decision?

"If your Daddy wants to stay here and drown, that's *his* ass!" Stephen retorted. "I'm lookin' after those kids and you!"

Unfortunately, like most men of that time and place, Stephen wasn't quite sure how to get even one child ready to travel, let alone seven, and when this fact dawned on him he acquiesced to waiting for Florence's parents to arrive. He piled blankets, food, and water jugs in the station wagon, then returned to the radio to listen for weather updates that never came.

Around 6:00 a.m., with the wind howling, Stephen met his wife's parents as they entered the driveway. Expecting his father-in-law to be an ally, he shouted, "Please go in there and explain to Florence that we need to leave *now*!"

The older man's response was a cruel surprise. "What are you, boy, a coward? I've seen blows like this that didn't amount to crap, and we'll ride out this one too. Right here." He punctuated his words by jabbing his finger toward the ground. To Stephen, a World War II veteran of the Pacific theater, the suggestion that he was a coward was like having a knife stuck in his ribs, then twisted.

Perhaps the outcome would not have been so tragic, Stephen would later reflect, if they'd had the luxury of another hour or so to discuss the matter. Unfortunately, within moments, water came pouring over the low bank in front of the house. Unbelievably, waves were breaking nearly five miles inland and, although they couldn't have known this at the time, at a point seventy miles from the eye of the hurricane. Stephen hustled his family into the station wagon, but the road was already flooded, and they got only as far as the end of the driveway before the water began to swamp the vehicle. With no alternative, they dashed back to the house.

The power was out, and they would get no more news reports. The wind drove the rain in sheets, and the house shuddered. Around 8:00 a.m., a mighty surge lifted the house off its blocks. As the home split and started to sink, the Broussards helped each other into the attic.

Struggling with her six-month-old daughter Estelle in her arms, Florence slipped and became wedged between a pair of joists. As she passed the screaming infant to her father, a wave crashed through the wrecked gable and swept the baby away. For Stephen it was a terrible choice: his daughter or his wife. He chose his wife first and success-fully freed her from the sinking framing. By then the baby was gone. The infant's body would never be found.

They climbed out on the roof, which itself was beginning to disin-tegrate. Florence held onto eighteen-month-old Michael. As the family locked arms to keep from being tossed into the water, they were bumped by the floating end wall of their ruined house. It seemed a better raft than they were on, and the grandparents and three of the children climbed onto it before the wind drove it away. Florence didn't see what happened next, but Stephen did. A wave broke over that sorry platform and swept the older folks and the three children into the floodwaters. Florence's mother lost hold of three-year-old Larissa and saved herself only by grabbing a dangling electrical cable. Four-year-old Veronica was also gone. Carolyn, age six, and her grandparents somehow climbed back onto the floating wreckage. A full day would pass before Stephen and Florence would learn that at least those three had survived.

The remaining section of roof dwindled to a mere three-by-seven-foot raft—a tight squeeze for Florence, Stephen, and three children. Stephen snagged a quilt from the wind-whipped water and bound everyone together, then twisted the corners around his own arms and held onto them with his full strength. Through the rest of the day and into the night, the rain and waves beat them relentlessly. The hundred-mile-per-hour wind swept them into an open lake. The rain struck them like bullets, hour after hour.

The inland waters in these parts teem with creatures that shun the salt water, which was now flooding their habitats. After nightfall, Stephen felt snakes crawling up his arms and legs. Each time one slithered onto him, he grabbed it and pitched it back into the darkness. Then Stephen Jr. complained that something had bitten him. Although the elder Stephen tried to assure everyone it was just a water bug bite, the reality became unavoidable when the boy began to hallucinate. They were helpless to do anything but stay clustered together and hang onto their feeble raft.

It was a long and miserable night—with three children almost certainly drowned, the fate of another and Florence's parents unknown, and Stephen Jr. with his snakebite. At daybreak, two pairs of fang marks stood out unmistakably on the child's swollen right cheek and ear.

The storm had driven them more than two miles across White Lake. Recognizing a hunting camp on shore, Stephen plunged in and swam through the debris to fetch fresh water, praying that no snakes would bite his own face when he surfaced for air. As he filled a kettle from the hand pump in a battered cabin, he noticed that two soggy bunk beds were curiously dark. On closer inspection, he found them covered with snakes.

Then he heard a boat motor.

Arriving to check on the damage at his hunting camp, the owner ferried the Broussards to the south landing. Unfortunately, it was too late to save Stephen Jr., and the boy died en route to the Abbeville hospital. Stephen Sr. spent the next three weeks unsuccessfully searching for his three missing daughters. He and Florence had lost four of their eight children plus their home and all of their belongings in that one terrible twenty-four-hour period.

"Nothing I experienced in three years of war in the Pacific compared to that night in the hurricane," he says. With tears in his eyes and raw pain in his voice, he adds, "I still think about those children every night." Every night for forty-five years.

After making eye contact with Florence in a nonverbal communication that only the two of them understand, he goes on. "I also stopped drinking. Not immediately, but a number of years later. And then I saw a doctor who prescribed some pills for my depression." That's something he hasn't told many people, he explains, but it's something that might be relevant. And there's something else. He never managed to forgive his father-in-law, a man who to his death demonstrated little contrition for his contribution to the demise of those four children. Tolerance for his father-in-law flirted with its limit a few years after Audrey when the man accidentally cut off two of Stephen's fingers with a power tool.

Yet even this was not the end of the Broussards' misfortunes. A few years after the hurricane, another son was killed by a car while riding his bicycle. Then, just a few years ago, an adult son, Brent, died in the crash of a small plane. It is the reflection on Brent's death, the most recent, that comes close to cracking the veneer of Florence's composure.

How did their marriage manage to survive such an astounding string of tragedies? Without hesitation, Stephen points at Florence. "I love that woman!" he says. Glassy-eyed, she returns him a loving smile.

Although the Broussards briefly considered relocating farther inland after the 1957 disaster, they ultimately decided to rebuild on the site of their earlier home with the mutual commitment that they would leave immediately at the first warning of any hurricane. Florence gave birth to another five children, for a total of fourteen. She named two of them after their siblings who had died in Audrey.

Hurricane Audrey was a disaster on multiple levels. In the parish seat of Cameron alone, the storm surge drowned 395 people—more than one-third of that little town's population at the time. More than another 100 died elsewhere in Cameron Parish and in lower Vermilion Parish, and hundreds who did survive incurred serious injuries ranging from fractures to snakebites. Virtually everyone in that huge region lost one or more loved ones in the hurricane, and several entire families were wiped out. At least 102 bodies were never found. The flooding extended as far as twenty-five miles inland along seventy miles of shoreline, sweeping away the homes of people who never thought of themselves as living particularly close to the coast. No structure in the region escaped damage. Five thousand people—the vast majority of the residents of that place at that time—were left homeless.

The dead, injured, and the dispossessed were victims not just of a natural phenomenon but also of an unfortunate confluence of human mistakes and oversights. Not only had the Weather Bureau's early warnings understated the danger, but inexperienced announcers at small rural radio stations delivered those advisories lackadaisically and sometimes inaccurately. Locally, there were no public emergency plans in place. Even if such plans had existed, and the residents had been aware of them, they would have been useless unless some knowledgeable authority had contacted the local public officials directly. As it was, the local officials received no better advice or information than their constituents got directly from their own radios—information that ran late by six to twelve hours.

The catastrophe, of course, did nothing to enhance these rural folks' confidence in the distant federal bureaucracy that issued national hurricane forecasts. Even as the oil companies had been making correct and timely decisions to safeguard and evacuate their offshore platforms, the common citizens had been erroneously assured that they had another full day to prepare before Audrey's landfall. In the aftermath, many victims vilified the meteorologists as inept, overpaid, unfamiliar with the regions their forecasts affected, and even lacking the courage to contritely visit the places where people had died because of their ineptitude. In Cameron Parish, a group of angry survivors banded together to bring a class-action lawsuit against the Weather Bureau and the radio stations. But, nine days after the disaster, every tape and transcript of the local stations' weather announcements mysteriously disappeared. Lacking the necessary evidence, and in the face of the Weather Bureau's virtually impregnable claim of legal immunity, the lawsuits never progressed.

The forecasters at the Weather Bureau were the only group that saw a meager upside to the catastrophe. If such forecasting and communication errors had been made in the face of a similar fast-moving and accelerating hurricane heading toward New Orleans and the eastern coast of Louisiana, several hundred thousand people would have been trapped and the death toll might have run into the tens of thousands. Eventually, someday, a hurricane like that was bound to come along. And when that happened, hopefully the whole central Gulf Coast would remember at least one two-word lesson from Audrey: storm surge.

CHAPTER 3

BAYOU COUNTRY

Cultural geography is never disconnected from physical geography, and that's especially true for Louisiana. Roughly one-sixth of the Pelican State's total area—nearly eight thousand square miles—is covered by water, with most of this submerged land lying in the southern parishes. The state's rainfall averages sixty-four inches a year, vying for the highest in the continental United States. Although the winter rains are sometimes daylong drizzles, the summer showers tend to come in compact squalls and thunderstorms that dump torrents as they race through, sometimes drenching one side of a community while leaving the other side bone dry. And, of course, the state also gets its share of the ultimate rainstorms: hurricanes.

Just east of the region devastated by Audrey sprawls the territory immortalized by Longfellow's epic poem *Evangeline,* the tragic story of lovers parted when the British deported thousands of French-speaking people from the Acadian region of Canada in 1755. Most of the displaced Acadians (the word eventually evolving into "Cajuns") settled around the bayous and swamps of southern Louisiana, where virtually nobody else would have considered living. Here, isolated from the political turmoil of the rest of the world, they learned to thrive in peaceful harmony with nature.

The soggy soil supported squash, okra, and tomatoes, which the Cajuns stewed in gumbos with alligator meat, crawfish, and catfish. Rice also thrived, and mixing it with the other edibles led to the dish they called "jambalaya." The more energetic fishermen poled their pirogues down the waterways to the Gulf, where they scooped up abundant shrimp and oysters. Unlike in the eastern provinces of Canada, with their bitter winters and rocky soil, food in southern

Louisiana was plentiful year-round. And although the Cajun homes were small and simple, they were built of cypress, which was impervious to rot. This was a place where families could literally live from day to day, oblivious to the insanity of the outside world. The pace of the Cajun lifestyle adjusted itself accordingly.

Being born in southern Louisiana was hardly different from being planted there, and even until recently many inhabitants blissfully lived their entire lives without traveling beyond the next parish or two. When other parts of the South were mired in racial discord, here musical groups of mixed color gave birth to zydeco, that particular combination of French and African music set to a polka rhythm and played on guitar, accordion, and washboard. Saturday mornings were, and still are, a time of revelry in small towns in Acadiana, with restaurants serving beer at breakfast as their dance floors swarm with Cajuns, Creoles of mixed color, and the occasional Latino or African American cowboy. The majority are reverent Roman Catholics who seem to figure that Christ would never have turned water into wine had he not intended to establish an important priority for his followers.

Although some of Acadiana's isolation had frayed by the mid-twentieth century, the fundamental culture had not. Yes, as of 1934 there was an intracoastal waterway slicing westward from the Mississippi River to the Texas border. Motorized shrimp boats replaced most of the pirogues, and shrimp was actually being exported. And by the late 1950s, every road south seemed to terminate at a canal clustered with oil-drilling equipment ready to be towed offshore or having just been brought in. But by outside standards, traffic on the Louisiana intracoastal canal was relatively sparse; all of the oil equipment clustered at the south ends of the roads reflected the simple fact that there just weren't very many roads, and the shrimp boats were hardly harbingers of an industrial age.

Well into the 1970s, television reception in the area amounted to a few snowy-screened channels from New Orleans and Lafayette, which for many Cajun families didn't justify the expense of a TV set. Radio was the main conduit for getting news from the outside world. Although admitted to the Union as the eighteenth state back in 1812, Louisiana never did bother to replace the Napoleonic Code as the basis of its legal system, nor did it ever start referring to its political subdivisions as "counties" rather than "parishes." Louisiana was what it was, and its people and its politicians liked it that way.

Even in the latter third of the twentieth century, for many folks, owning a boat was a higher priority than buying a car, and with relatively few outsiders having business here, there was no good reason to pave many of the roads or to erect road signs. In fact, road signage still remains a future agenda in much of this region. During the Civil War, Union troops learned to their chagrin that overland campaigns here were difficult or impossible, and even today some would argue that ground travel hasn't become a lot easier.

There are but a handful of places where the Louisiana coast is accessible by car—a few dozen miles near the Texas border, several obscure boat launches, the gated supertanker terminal at Port Fourchon, and adjacent Grand Isle in the east. In fact, to drive most places south of the east-west State Route 14 still requires a commitment to an adventure on long, narrow, poorly marked, and often unpaved roads that never take the shortest path between any two points. Regardless of where you start and where you're going, you can be guaranteed that your route will wiggle around a labyrinth of swamps, bayous, and lakes—provided, that is, that you don't get onto one of the many roads that simply end at the edge of a swamp in the middle of nowhere.

Even assuming that you don't get lost (and virtually every outsider does), your odometer will run up a mileage far out of proportion to the scale of the map you're following. The straight-line distance from Cameron to Lake Charles, for instance, is 27 miles, yet the two land routes measure 57 and 50 miles, the latter involving a ferry. The lengthiest compulsory detours are in the southeastern parishes. Grand Isle is only 23 miles from Port Sulphur as the crow flies, but by car it's a whopping 189 miles, all of which are west of the Mississippi River.

The southern parishes span an expansive tidewater grassland gouged with thousands of shallow channels and lakes that teem with millions of alligators, turtles, nutria, water moccasins, and catfish; billions of mudbugs (crawfish); and a wide assortment of birds. At the upper border of Acadiana, a hundred miles north of Vermilion Bay, the great silt-laden Mississippi River splits into two branches. The smaller and more direct western river, the Atchafalaya, flows through a series of shallow lakes that empty into a muddy bay, while the serpentine eastern branch, the Mississippi, is contained by levees so that it flows safely past, and not into, the city of New Orleans. Because these two major rivers are higher in elevation than most of the surrounding land, the runoff from local rainfall flows *away* from them, not into them, on its way to the Gulf.

In this region, the land is so low that much of it alternates between being dry and being underwater. With such little variation in elevation, the myriad streams are sluggish and sinuous, and in some of these, the bayous, the direction of flow actually reverses from time to time. There are no beautiful beaches along these shores of the Gulf, for what settles from the many streams is mud, not beach sand. All of that sediment, however, is chock-full of nutrients that support a vast and complex ecosystem of grasses and algae, mosquitoes and beetles, fish and fowl, arthropods and reptiles. Plus, on the other end of the food chain are thousands of hunters, fishers, and oil-field workers.

In the 50 years prior to Camille, fourteen hurricanes and twenty-seven tropical storms made landfall in Louisiana—an average of one hurricane every 3.6 years and one tropical storm every 1.9 years. Such arithmetic does not, however, mean that all of the residents were personally experienced with hurricanes. Because these storms were scattered along a 397-mile coastline, hurricanes striking the western part of the state might have but marginal effects in the east and vice versa. Moreover, because very few people lived within a few dozen miles of the Gulf, and even in the bayous no hurricane can sustain devastating winds for very long after it makes landfall, most folks' direct experiences with tropical storms were likened to those of a very bad thunderstorm. And summer thunderstorms, of course, were something everyone was accustomed to.

Only after 1957 did residents of southern Louisiana begin to respect hurricanes as serious threats to life and property. Today's heightened level of hurricane awareness in this region owes much to the terrors of Hurricane Audrey, not because many of today's residents are old enough to remember that catastrophe, and not because many have even read about it (in fact, very little has been published about Audrey), but rather because of the strength of oral family traditions. In a region where few move in and fewer move out, virtually everyone living here today has kinfolk who tell gut-wrenching stories about the 1957 disaster—stories that in their telling and retelling make indelible impressions on the next generation.

Although official state maps include New Orleans as part of Acadiana, to most Louisiana natives the Big Easy might just as well be in another country. As far as the rest of the state is concerned, Plaquemines Parish, which straddles most of the hundred miles of the Mississippi River between New Orleans and the Gulf of Mexico, might just as well be on another planet.

In 1699, Pierre Iberville established the first seat of government for the Louisiana territory near present-day Biloxi, Mississippi (pronounced "B'lucksy"). Then, searching for a better deepwater harbor, Iberville moved the capital to the present site of Mobile, Alabama, in 1710. Yet it was clear that all of these early sites were too far east to secure the French claims over the Mississippi River watershed. What the French needed was a city on the banks of the great river itself.

Exploring and charting the river's mouth was a more formidable task than anyone anticipated. To everyone's surprise, it was not a single channel but rather a confusing labyrinth of shifting sandbars and islets, tidal swamps, meandering estuaries, and weird "mud lumps" that sometimes arose overnight and made a previously navigable channel impassable. The shortest and most direct way in, it turned out, required a ship to enter from the *northeast.*

The solid land closest to the river's mouth was not along the river at all but on the coast of the Gulf of Mexico about eighty miles north of the entrance to the Main Pass, on land very close to Iberville's original coastal settlement of 1699. Accordingly, in 1719, the capital of the Louisiana territory was moved west once again from Mobile back to today's Biloxi.

With the assistance of the local Indians, Iberville and his settlers continued to learn more about the local geography. And a curious geography it was. If you followed the coast west from Biloxi for about fifty miles, paddled into an immense lagoon, beached your pirogue on the south shore, then hiked about six miles through a shallow bowl of reasonably solid ground, you would come to an upslope that turned out to be a natural levee of the Mississippi River, at a point 107 miles upstream of its mouth.

The French company led by Iberville's younger brother, Jeanne Baptiste Bienville, recognized not only that the land between the river and the lagoon was a potential site for a town but that this was the southernmost practical site for a full-sized city before the marshlands began. Bienville named the place New Orleans. It was, in the words of one writer, "an inevitable city in an impossible location." Today, it is the only city in the United States with an average elevation below sea level.

In 1718, when the town's construction had barely begun, the river overflowed its natural levee and submerged the entire project under a foot of water. Clearly, this couldn't be allowed to happen very often, and when the water evaporated, the first order of business was to build a dike three feet high to prevent a recurrence of such an event. Then in 1722, a hurricane sideswiped the city, and the river surged nearly eight feet, topping the paltry dikes. Clearly, the levees needed to be stronger, higher, and more extensive. And, indeed, each subsequent flood through the following centuries provoked New Orleanians to make further improvements to their levee system.

The quality of building construction had improved since the early years, and with the rebuilding after the 1812 hurricane came increased attention to issues of structural integrity. Moreover, with Louisiana's admission to U.S. statehood that year and the designation of New Orleans as the state's first capital, financial resources became available to further expand the levee system and to install the first crude pumping stations to lift storm water out of the city. Hurricanes and tropical storms continued as they always had, but the improved infrastructure decreased the frequency of damaging events.

Situated so far inland, New Orleans was not particularly threatened by extreme winds; the major concern since the earliest days had always been the prospect of flooding. By the mid-twentieth century, the historical record seemed to demonstrate that the extensive and ever-expanding levee system was indeed effective in holding both the river and Lake Pontchartrain out of the city. In the aftermath of the great flood of 1927, which inundated the homes of nearly a million people in the river's upstream watershed but spared New Orleans, the Army Corps of Engineers took over the maintenance of the levees and embarked on a major project of improving and extending them even farther. And although the levees superficially appear to be mere piles of dirt, in fact they are carefully designed feats of engineering. By the 1950s, there were a thousand miles of levees in southern and eastern Louisiana, and at New Orleans some dikes stood seventeen feet above the adjacent streets. Downstream, artificial riverbanks extended more than ninety miles south of the city.

Unfortunately, this "man over nature" strategy has had its unanticipated consequences. By depriving the great river of its natural floodplains, each expansion of the levee system has actually increased the vulnerability of other parts of the region to flooding. After all, when the Mississippi rose, whether by virtue of rains in the heartland or the block of its egress by a storm surge in the Gulf, the unrelentless flow had to go *somewhere*.

Meanwhile, New Orleans was sinking. Centuries before human intervention, the common floods regularly deposited silt over the floodplains, replacing land that had eroded or subsided. With the river contained between dikes, this natural renewal was halted; as a result, some 85 million tons per year of waterborne silt was being forced out over the edge of the continental shelf, where, instead of rejuvenating the land, the sediment tumbled into the depths of the Gulf of Mexico. Pumping water from the ground aggravated the process, nearby oil and gas extractions didn't help matters, and some sections of the city sank to nearly ten feet below sea level. To the south, as hundreds of square miles of low-lying lands gradually disappeared, the Gulf crept slowly toward the Big Easy. New Orleans became a bowl surrounded by dikes.

Although no large city is ever easy to evacuate, New Orleans presents a particularly nightmarish challenge. Today's emergency planners figure that by funneling everyone into the current four-lane highways, all of which cross waterways, a complete evacuation would take almost three days. Thirty-five years ago, at the time of Camille, only one interstate highway and a few low-lying two-lane roads led out. Planners then didn't even consider the prospect of evacuating the city. They knew, a priori, that such a notion was out of the question.

Create a place, and there will always be people who figure out how to live and thrive there. So it was with Plaquemines Parish, only one-third of which is dry land. This hundred-mile-long peninsula, bisected by the Mississippi River and created by many thousands of years of accumulated silt, became home to a mixture of Croatians, Italians, Spaniards, and other European immigrants, plus a smattering of blacks.

One village, aptly named Pilot Town (also spelled Pilottown), sprang up at the "Head of Passes" at mile zero of the Mississippi River, south of which the official river mileage runs into negative numbers. This is where bar pilots disembark from incoming freighters to be replaced by river pilots who guide those ships farther upstream to New Orleans or Baton Rouge, with the exchange reversing for downstream-traveling shipping. There has never been a road to Pilot Town, nor was there one to the several other villages, now long gone, that sprang up in these parts. There are spots on this planet where roads just don't make sense, and much of Plaquemines Parish is still one such place.

Except for the few families in Pilot Town with occupational ties to the upstream cities, the early residents of Plaquemines Parish and the New Orleanians generally considered each other mutually irrelevant. In fact, there wasn't so much as a single road connecting any part of Plaquemines Parish with *any* other parish until after 1900. Then in 1927, after an astonishing act of territorial aggression spearheaded by the business community in New Orleans, Plaquemines Parish residents ideologically solidified their historical isolation from the people and politics of the rest of the state and even from the rest of the nation.

That year, 1927, the spring rains in the nation's heartland were relentless, and the city of New Orleans was being threatened by a great flood that had already burst one upstream levee after another, rendering a million people homeless in Illinois, Tennessee, Arkansas, Mississippi, and northern Louisiana. Yet the upstream floods were no relief to those downstream, because ultimately all of the floodwaters still drained back into the Mississippi basin to continue their unrelenting journey to the Gulf. In New Orleans, the river level flirted with the tops of the levees.

A group of fifty-one prominent bankers and businessmen decided they could do something about the problem. All they had to do was to destroy the levee just downstream of the city: open a big enough drain down there, and the giant pool would be flushed. No matter that it would be flushed into Plaquemines and St. Bernard Parishes; the important thing was to save New Orleans. Those business leaders brought their influence to bear on the politicians, both state and federal, and the powers quickly approved.

The district attorney and former judge for the flooded parishes, Leander Perez, was furious that outside interests and self-serving politicians could consider wreaking such hardship on his constituents. First he shouted and demanded, then he tried to negotiate, and ultimately he pleaded, only to be forced into a humiliating compromise on April 26, 1927. The citizens of the two southern parishes would sacrifice their homes on the assurance—ultimately false—that they would be adequately compensated.

Never again would Leander Perez respect or trust outsiders.

On April 27 and 28, 1927, under orders from Louisiana governor O. H. Simpson and with the assistance of the Louisiana National Guard, the ten thousand residents east of the river in Plaquemines and St. Bernard Parishes were forcibly evacuated. The following day, while troops stood with weapons poised to deter any interference, state workers dynamited the east levee at Caernarvon, thirteen miles below New Orleans. As planned, the Mississippi River surged through the crevasse. Several thousand homes, farms, and small businesses were destroyed. The involuntary refugees lost almost everything they owned in the intentional destruction, and despite court battles that dragged on for the next few years, their final average compensation from the city and the state was about ten cents on the dollar. The state courts ruled that the entire egregious episode was perfectly legal.

As it turned out, the intentional flooding of Plaquemines and St. Bernard Parishes was as unnecessary as it was unconscionable. By the time the dynamite was detonated, the river level had already begun to drop, and a significant fraction of the floodwaters had been unexpectedly carried off toward the Gulf of Mexico by the Atchafalaya River. But human decisions sometimes have a momentum of their own even in the face of new information, and with photographers present, armed troopers standing guard, and curiosity seekers and reporters circling overhead in biplanes, the levee had been blasted nonetheless.

The simplest of people are often the most resilient, and virtually everyone displaced chose to return and rebuild on their original homesites. There was, however, now a big change in the local political culture in Plaquemines and St. Bernard Parishes. After the egregious act of homeland terrorism by that city's captains, New Orleans would no longer be viewed as irrelevant by the folks downstream. Instead, the city's leaders would forever be considered a potential threat to life and livelihood. The same distrust would apply to the state legislature and the entire court system of Louisiana. And despised just as much by those in the sacrificed region was the federal government, which had permitted the intentional flooding to proceed.

CHAPTER 4

THE BIRDSFOOT PENINSULA

Louisiana has been graced with a passel of colorfully corrupt politicians, including the likes of former governor and U.S. senator Huey Long, who was assassinated in the extravagant state capitol building he built; the recent governor Edwin Edwards, who as of this writing is serving a ten-year term in federal prison for bribery and racketeering; and three consecutive insurance commissioners who were convicted of various betrayals of the public trust. It was Louisiana that gave the nation David Duke, the boyish-looking grand wizard of the Ku Klux Klan and one-term state legislator who unsuccessfully ran for governor, U.S. senator, and twice for president and who went to federal prison for bribery and racketeering. But it was the shenanigans of Leander Perez—a one-term judge and multiple-term parish district attorney—that set a record for political malfeasance that few are ever likely to match. And it was largely through the political brawn of the same Judge Perez that Plaquemines Parish would be prepared for Camille, losing only nine lives out of some twenty thousand residents at risk.

Born in 1891, Perez grew up in Plaquemines Parish at a time when it lacked any roads connecting it with anywhere else and the only ways to leave the parish were by boat or on horseback on the levee. With the encouragement and support of an uncle, he attended Louisiana State University and then earned a law degree from Tulane University. After earning a disappointing $350 during his first year as an attorney in New Orleans, Perez returned to Plaquemines Parish to work in the Clerk of Courts office.

When in 1919 the parish's district judge drowned in a fishing accident, Leander's uncle pulled strings with the governor and got the

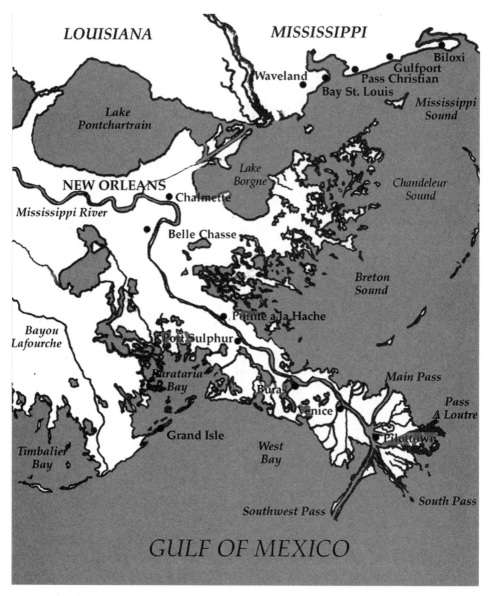

Plaquemines Parish and environs. (*Adapted from Louisiana Department of Transportation.*)

young attorney appointed to the bench for the rest of the unexpired term. Although the bench suited his temperament, Perez overstepped his bounds so quickly and so perniciously that in 1924 he was impeached for a long list of judicial misbehaviors. After several days of brutal testimony, the Louisiana Supreme Court recessed and worked out a closed-door agreement with the principals. The prosecution withdrew its complaints, after which the adversaries made stiff but polite public apologies to each other. A few months later, Judge Perez resigned from the bench to run for district attorney of Plaquemines and St. Bernard Parishes. He would nevertheless continue to use the title "Judge" for the rest of his life.

Perez had meantime forged a clandestine alliance with a financially successful smuggling ring that was bringing in foreign liquor through the swamps and trucking it via the back way to New Orleans. The bootleggers were pleased to have the friendship of a judge, and Perez was reciprocally pleased by those boys' good manners. Money flowed into Perez's pockets—money that he would use wisely.

In Louisiana, the office of district attorney is inherently more powerful than that of district judge. An unscrupulous district attorney can choose whom to arrest and for what reasons, he can decline to prosecute those he favors, and he acts as legal counsel for every public agency and board of directors in his parish—some of which handle lots of money. With his payoffs from the liquor smuggling, Judge Perez could afford to buy all the votes he needed, and in 1925 he won the election for district attorney by a landslide.

If Perez was a fool at first, he didn't remain so for very long. He soon realized that to establish and maintain his power base he not only had to look over his own shoulder at every minute but also needed to address the needs of the common folk (or at least the white common folk) of Plaquemines Parish. He quickly gained a reputation for being as generous to some as he was brutal to others. He paid from his own pocket for promising students to go off to college, yet he demanded an undated letter of resignation from everyone he hired.

Perez established two free ferries so folks who didn't own a boat could cross the river (even today these two ferries are still the only way to take a vehicle across the Mississippi anywhere south of New Orleans), yet anyone who angered him risked being tossed in jail on trumped-up charges. He prevented more blacks from moving into the parish by requiring outsiders to register for "work permits," and he prevented those blacks who were already there from registering to vote. He just didn't like colored folk, and that was that.

The debacle of the intentional flooding of 1927 cemented Perez's resolve that Plaquemines Parish could not play by others' rules. That fall, Perez helped elect Huey Long to the governorship by delivering him more votes than the number of voters in the parish. Yet in return, Perez asked for no immediate favor other than to be left alone. Huey Long suggested that maybe he should just cut off Plaquemines Parish and let it drift into the Gulf. "I wish you could!" Perez retorted.

The following year, 1928, just as the levee boards were incurring huge debts for repairs and improvements after the great flood, oil was discovered in Plaquemines Parish. Over the next few years, a great stream of money gushed from the ground and flowed straight out of the parish. This was a challenge that called for an ingenious plan, and Perez thought of one. Although Huey Long resigned the governorship in 1931 to become a U.S. senator, he still controlled the state politically and he still owed Perez a favor. Perez called on him, and with Long's backing the state legislature passed a constitutional amendment allowing the Plaquemines Parish police jury (the official parish governing body) to take over the bonded indebtedness of any levee district located wholly or partly within that parish and to hold the assets of such levee districts *as long as they remained in debt.* About a third of the levee districts in Louisiana were at least partly in Plaquemines Parish. And all of them owned extensive mineral rights.

Mineral royalties from oil, gas, and sulphur immediately poured into the parish coffers, the windfall income far exceeding the small amounts the parish had to return to the levee boards to pay the interest on their bonds. And, as their legal counsel, Perez had the power to keep the levee boards in perpetual debt by periodically issuing new bonds and by manipulating the appointments of the levee board commissioners (whose signed undated resignations he kept in his locked file cabinet). Perez also had friends in the State Land Office, who, at his request, acquired federal swamplands whose ownership was then transferred to the levee boards. By manipulating school board elections, he also gained control of that body and thereby control over the mineral wealth that lay beneath school district property.

When he died in 1969, the year of Hurricane Camille, estimates of his net worth ran as high as $100 million, a remarkable achievement for a man whose public salary never exceeded $7,000 a year.

How did he acquire such a huge personal estate? Nobody has ever succeeded in explaining it all. What we do know is that Perez created dozens of dummy oil and mineral-leasing corporations run by his friends and relatives, that those businesses were in collusion with the levee boards, and that he charged outlandish legal fees to those puppet companies. Because most were incorporated in other states, the money trails were extremely difficult to trace.

Perez required every mineral leaser in the parish to hire a percentage of local workers and to lend equipment to the parish whenever it was needed. Soon the Plaquemines Parish unemployment rate was the lowest in the country. Meanwhile, Perez stiffened the ordinances requiring outsiders working in the parish to be photographed, to be fingerprinted, and to pay for work permits. Very few blacks ever seemed to quality for such permits. "We could easily have 100,000 people in this parish, but we don't want 'em," Perez explained. Indeed, the parish population never exceeded twenty-four thousand residents in his lifetime. It remains less than twenty-seven thousand today.

Elections were farces. It was not uncommon for a Perez opponent to be arrested on spurious charges, with bail set extraordinarily high and all authorized officials curiously unavailable to process the paperwork when the accused did come up with the bail. The transgressor would languish in jail for several days before being released, having become a bit smarter about trying to oppose "Judge" Perez again in the future.

Perez was diligent about getting out the vote in every election, using teams of paid volunteers to drive people to the polls and paying voters between two and ten dollars per vote. He was careful, though, not to transport the two-dollar voters in the same cars with the ten-dollar voters; people did, after all, have their pride. When Perez testified before a U.S. Senate committee about this practice, Senator Hugh Scott of Pennsylvania asked him, "You segregated them according to how much you paid them, then?" At Perez's unapologetically affirmative answer, Scott sarcastically referred to the practice as "the current Louisiana Purchase."

Businesses in Plaquemines Parish found it much easier to cooperate with Perez than to oppose him. They needed building permits, drilling permits, pipeline permits, and various licenses to do business—the granting of which were all controlled by Perez. Yet in many ways Perez was a benevolent despot. The parish had no local sales or income taxes, and property taxes were the lowest in the state. Businesses were required to submit an annual report on the man-hours of local labor they hired and of materials purchased in the parish, so Perez could make sure that locals were getting enough of the jobs. During his reign, the parish saw the construction of paved streets, libraries, more levees, water purification plants, public schools, and a modern hospital at Port Sulphur. He began garbage collection services and a mosquito control program and had recreation grounds set aside and parks built. As early as the 1950s, he had a parish-wide emergency communication system. He instituted health and life insurance programs for all parish employees and started a college scholarship program. In addition, he provided legal work for the indigent without charge and loaned money to the needy.

Tropical storms are a part of life in Plaquemines Parish. With every tempest, Perez galvanized the locals into action even as he summarily refused outside help. After Hurricane Flossie in 1956, he was highly visible at the courthouse for the next thirty days, handing out food and clothing and making sure that everyone received needed medical attention. Including blacks. Not that he'd changed his mind about liking black people. It was just something a good Catholic was supposed to do under such circumstances.

If Plaquemines Parish had any rock outcroppings (it doesn't), Perez might have tried to have his political philosophy chiseled in stone for posterity. He vehemently opposed racial equality, he opposed federal control of oil and mineral resources, he opposed national welfare and public works programs (he saw no contradiction between this view and his own paternalistic actions on the parish level), and he opposed even the existence of labor unions. The federal government's endorsement of all of these ideals—plus its historically irresponsible inactions during the parish's intentional flooding in 1927—caused Perez to hurl constant criticisms at Washington. He detested the very concept of a centralized U.S. government. As for his international perspective, he was fiercely opposed to the United Nations, believing it to be but one step away from a world government that would be ruled from Moscow.

Perez explained at every opportunity that blacks are inherently inferior to whites, both intellectually and morally. He stated in public that there are two kinds of Negro: "Bad ones are niggers and good ones are darkies." After the *Brown v. Board of Education* Supreme Court decision in 1954, Perez gained national notoriety when he took his campaign to preserve segregation on the road. His bombastic message was this: The movement to racially integrate schools and workplaces was a scheme to undermine the racial purity of whites by mixing the races and thereby to destroy America's moral fiber. This plot of genetic sabotage was not designed by the Negro himself, who was intellectually incapable of such an elaborate program, but rather guided from the Kremlin. The blacks were pawns of the communists, and every white man who preached integration obviously took his orders directly or indirectly from Moscow. If the civil rights movement were successful, the Negroes in the South would be given the right to vote, and they would be told how to vote by the communists. This would destroy the South first, because blacks were in the majority in some southern towns and counties. White schools would be overrun by intellectually inferior black children, educational standards would be lowered, and within a generation America's economic and technological superiority would be an artifact of history. The Soviets would thereby succeed in destroying America without dropping a single bomb.

Given that the Negroes were merely the ignorant dupes in this unfolding disaster, who were the real agents of the communist conspiracy? Perez loved being asked this question, for he had the answer. The Zionist Jews were behind it all. Everyone, after all, *knew* that the National Association for the Advancement of Colored People (NAACP) was a communist front run by Jews. The national political parties, both Democrat and Republican, had sold out to these traitors because the Jews would be able to deliver the Negro vote as a bloc. Of course, this message was difficult to get out to the public because, as everyone well knows, the media is controlled by Jews. But that wasn't all. There was now a crazy movement to admit Hawaii as a state, which would make matters even worse for the defenders of American liberty by including mixed-race, left-leaning representatives of that remote Pacific island in the U.S. Congress.

Perez began traveling to Washington to testify before congressional committees and to otherwise oppose proposed civil rights legislation. Although he found a number of allies, including James O. Eastland, U.S. senator from neighboring Mississippi, the Civil Rights Act of 1957 became law nonetheless. Henceforth, the federal government was empowered to bring suit in federal courts to prevent voter discrimination.

Between 1936 and 1953, not a single black person had registered to vote in Plaquemines Parish. Although Perez claimed that this remarkable statistic was due to blacks' lack of civic responsibility and had nothing to do with discrimination, in fact, the parish's registrar of voters was well known to be an elusive figure who mysteriously disappeared when any black resident showed up to register. Matters improved only slowly after 1957; in 1962, there were 6,906 whites and 43 blacks registered to vote. This was out of a population of 22,545, of which almost 29 percent were black. The culture of disenfranchisement continued to run deep: most blacks knew the risks of trying to register—losing a job, being ostracized socially, receiving higher tax assessments, and/or losing credit at local stores. Such repercussions would be a high price to pay for casting a vote that would probably not be counted anyway.

Unlike in Louisiana's Cajun parishes, where race relations were generally amicable, racism was pervasive in Plaquemines Parish. When Perez hosted a group of students from Sarah Lawrence College who were studying life in the South, he explained that the reason for separate black and white water fountains was that almost all Negroes have communicable diseases. He withdrew from a Red Cross hurricane shelter program because that agency would not guarantee that the shelters would remain segregated. And he insisted that the parish health department withdraw from the Louisiana Department of Health because that agency had accepted federal funds. He was taking no chances of having the feds force the desegregation of health services in his parish.

When the NAACP, the Congress for Racial Equality (CORE), and other civil rights groups became active in the South, they sponsored no activities whatsoever in Plaquemines Parish. Perez, rather than being pleased that they were staying away from his fiefdom, became suspicious that some clandestine preparations were in progress. Then in October 1963, several groups of northern activists showed up in the small town of Plaquemine, which is in Iberville Parish, more than a hundred miles northwest of the Plaquemines Parish line. Perez quickly convinced himself that the folks running CORE had actually intended to invade his own parish but, due to their ignorance of the regional geography, ended up in the wrong place. Accordingly, he made preparations to deal with any outside agitators who might try to enter his domain in the future.

Deep in the swamps east of the Mississippi River stood the ruins of a garrison named Fort St. Philip, built in 1724 and now privately owned. Surrounded by marsh, it could be reached only by boat or helicopter. Water moccasins, nasty-tempered spiders, and swarms of bloodthirsty mosquitoes, coupled with the heat and humidity, made this the perfect place to incarcerate any civil rights demonstrators or pesky northern news reporters who might come around making trouble.

The parish council rubber-stamped Perez's plan to lease the place from its private owner. Perez announced to the media that he would fully integrate the new prison—jailing white and black anarchists together. Parish work crews strung an electrically charged eight-foot-high barbed wire fence atop the foot-thick, five-foot-high brick wall that surrounded the four-acre enclosure. The damp, sweltering underground rooms with bare masonry floors and six-foot-high ceilings would provide protection from the elements but not from the snakes, spiders, or mosquitoes. To try to escape would be futile, if not suicidal, since the fortress was surrounded by a moat twenty-five feet wide that was freely accessible to the local alligators. Beyond the moat on one side was the mile-wide Mississippi River, and on the other side were swamps as far as the eye could see. Demonstrators were to be transported to the fort aboard a forty-nine-foot converted trawler equipped with cattle pens. While transportation to the fortress would be free, the prisoners at Fort St. Philip would be required to pay for their own food and lodging while incarcerated, so as not to become an economic burden to the good people of Plaquemines Parish.

Nationally televised news programs quickly picked up the story of

Leander Perez's new civil rights jail. As Perez was filmed walking around the converted old fort, he promised, "A few days here, and anarchists will surely come to understand our way of life. This is a peaceful spot, a perfect place to meditate."

As it would turn out, nobody was ever jailed in the place. No civil rights organizer or demonstrator even entered the parish. That, quite possibly, is what Perez had planned all along.

When, in 1962, the Catholic Diocese of New Orleans announced its decision to desegregate that city's numerous Catholic schools, Perez raised such a ruckus that the archbishop excommunicated him. Perez had a suspicion of what was coming next, and he was right. Plaquemines Parish had a Catholic elementary school in Buras—Our Lady of Good Harbor—and the archbishop decreed that it too would be desegregated. Perez whipped his constituents into a frenzy, telling them that the outside efforts to integrate their local schools constituted a greater threat than the many hurricanes they had all experienced. Parents of children enrolled in the school began to receive harassing and even threatening phone calls, warning them to withdraw their children.

On the first day of the new school year, the two buses earmarked for parochial school transportation mysteriously developed mechanical problems, and only five black and thirty-eight white children showed up for classes. That same day, several of those pupils' defiant parents lost their jobs, while others had their businesses threatened with closure for various obscure regulatory noncompliances. On the second day of classes, no black children attended and the white student census was down to twenty-six. That evening, the priest phoned the press to announce that the school would be closed because of the lack of police protection in the face of threats of physical violence against the nuns and lay teachers.

The New Orleans archbishop, however, was not about to be bullied. He ordered that Our Lady of Good Harbor school remain open, if only as a symbolic gesture. Children would not be admitted this academic year, but for the next 249 days the teachers would show up daily at the empty school. For months, pickets continued in front of the building in spite of the fact that no teaching was taking place.

Shortly before the beginning of the 1963 school year, when the teachers, priest, and archbishop expected matters to return to normal, one of the nuns received a phone threat that the school would be blown to bits with her in it if they allowed black students in. That night, someone soaked the building's roof with gasoline and lit a fuse. The explosion and fire demolished one classroom and damaged the rest of the building. The next day, the archbishop announced that the school would remain closed for another year to protect the lives of all concerned.

In September 1963, Dan Rather narrated an hour-long program on CBS titled "The Priest and the Politician," which detailed the struggle surrounding the integration of Our Lady of Good Harbor. In the following months, the damage to the school was repaired. But Plaquemines Parish officials refused to grant an occupancy permit, and without that the power company could not reactivate the electrical service. Perez had won the battle. The Catholic school never reopened while Perez remained alive.

Also in 1963, Perez directed the parish school board to reject all federal funds for education. "Our children are not for sale for any filthy, tainted Federal bribes," he proclaimed. When pressed further, he rationalized that he was being patriotic in refusing such monies, given that the federal government was billions of dollars in debt.

The Johnson administration in Washington, however, was relentless in its efforts to end school segregation, and in 1965 the Department of the Interior threatened to withhold Plaquemines Parish's share of the oil royalties on federal leases in the Delta National Wildlife Refuge unless the parish complied with the Civil Rights Act of 1964. In 1966, Washington informed the parish school superintendent that he was obligated to devise a desegregation plan. When the Justice Department followed up on the continuing noncompliance by filing suit in federal court, Perez again swung into action. Vowing to fight "to the last ditch, come hell or high water," he unleashed his whole bag of intimidation tactics, legal subterfuge, and other shenanigans against the lawsuit.

Judge Christenberry, who had crossed paths with Perez many times over the years, heard the case. Despite all of Perez's legal maneuvering, Christenberry ruled in favor of the Justice Department and ordered the parish to comply with the desegregation order. Perez was livid. Christenberry's ruling further provided that, if he heard of any parish officials intimidating individuals or trying to persuade businesses to pressure their employees into avoiding integrated schools, he would bring in the FBI to investigate and file charges of obstruction of justice.

Using some of his own funds and soliciting donations from others, Perez took a different tack—establishing a network of private schools in the parish. The public schools descended into chaos, and it wasn't until 1971, two years after Perez's death, that they began to recover. Meanwhile, Perez continued his crusade of white supremacy and segregation on the state and national scene, where he was under no court injunction to keep his mouth shut. In this regard, Mary McGrory of the *Washington Post* wrote, "By comparison to Perez, Governor George Wallace of Alabama seems an angel of reason and moderation, and Ex-Governor Ross Barnett, of Mississippi, a towering intellect."

Attorney Luke Petrovich was one of Plaquemines Parish's political insiders for several decades. His father had immigrated from the Croatian section of Austria to become an oysterman in Olga, a village of four hundred (including just six U.S. citizens) that could be reached only by water. When that settlement was virtually obliterated by a series of tropical storms and floods, the family moved to Triumph, a community along the two-lane road west of the Mississippi River. It was there that, after Luke's father died, Judge Perez took notice of the young Croatian with mischievous blue eyes and a winsome smile, always working and ever generous about helping others. After checking into the boy's academic records, Perez reportedly offered to pay Luke's way through college, no strings attached.

After graduating and teaching for a while, Luke went on to law school. By the time he passed his bar exams in 1956, years had passed since he'd seen Judge Perez. At his swearing-in ceremony, it struck him as curious that a compatriot of Perez, a man named Kane, was sitting quietly in a back row. After the ceremony, Mr. Kane congratulated Luke and invited himself along for celebratory drinks with the new lawyers.

During the toasting and joking at the bar, Kane slipped in a question about Luke's employment plans. Luke answered that one of his prospects was to return to Plaquemines Parish to practice law with

Herman Schoenberger in Buras. Kane smiled. "Good idea," he said. "Do it. You won't regret it." Unbeknownst to Luke at that time, Leander Perez was planning to put his political machine behind a Schoenberger candidacy for parish sheriff in the future, and Luke was a prospective protégé to take over some of the parish's legal work.

A relationship developed between Perez and young Petrovich that was in many ways akin to that of a domineering father and an adopted son who is astute enough to see both the good and the bad in the old man. And smart enough to avoid endangering the relationship. Meanwhile, Perez apparently liked what he saw of Luke's work. In 1961, the local electorate approved a restructuring of the parish government from the then ten-member police jury to a five-member commissioner-council structure with Leander Perez presiding. Young Luke Petrovich was elected for a commissioner's post.

Ostensibly, the reduction in the number of parish officials was a move toward improved governmental efficiency and responsiveness to emergencies. Indeed, Galveston, Texas, had similarly changed its county government after its great disaster of 1900. But to Perez, there were also personal advantages to dealing with just one group of five commissioners who served both in executive and legislative capacities rather than ten men divided into two groups of five who might wrangle with each other. Plus, repaying five commissioners for special favors was certainly cheaper than rewarding ten.

This restructuring would eventually be reversed by a federal court one-man–one-vote ruling in 1986. For the intervening fifteen years, however, Plaquemines Parish was governed by a single group of five men who not only wrote the local law but also administered it. That period included 1969, the year Camille roared in from the Gulf.

As an indication of how patronage worked in Plaquemines Parish, on the day Luke was elected to the commission council a constituent congratulated him and, with a wink, asked what job he had planned for his brother-in-law. Luke, trying to think of the most outlandish possible response, laughed and said, "I think I'll make him snow-removal director." But Mother Nature can also have tricks up her sleeve, and that winter, for the first time in living memory, it actually snowed in Plaquemines Parish.

Each of the new commissioners carried a second title, and Luke's was director of public safety. Luke well understood that Perez, with his dictatorial yet paternalistically benevolent attitude toward the parish, expected him to take the responsibility seriously. And so Luke did, learning everything there was to know about emergency planning in 1961. To order an evacuation would require agreement by three of the five commissioners. And given the historical antipathy between Plaquemines Parish and outside authorities, such decisions would not be blindly based on advisories from the Weather Bureau (now the Weather Service), a federal agency whose nearest office lay considerably to the north in a suburb of New Orleans. Instead, the commissioners would rely mainly on radioed advisories from ships and the decisions of the oil companies that operated the offshore rigs.

Luke learned to lay circular cutouts over the nautical charts to develop scenarios for prospective evacuation decisions. He discovered, for instance, that if a storm 250 miles off the Mississippi coast had gale-force winds extending out 100 miles and was tracking north at 10 miles per hour, flooding could begin in southern Plaquemines Parish just eight hours later. If those eight hours happened to commence at nightfall, people would be trapped and drowning by morning, when the national weather folks would probably still be fussing around with their forecasts.

According to Luke's plan, the first people to evacuate would be the offshore oil workers. Next would be any of the shrimpers who didn't have time to run their boats up the bayous and canals. After that, residents at the high-risk south end of the road would leave while the folks farther north would stay put until after they passed. School buses would be pressed into service to transport families without cars. Every available truck would be loaded with oyster shells, ready to "shell a road" if portions of the narrow two-lane highway became impassable. If necessary, the shells would be spread along the top of the levee, and north-going traffic would be diverted up there. All of the firefighting equipment would be readied and manned in anticipation of the fires that were prone to break out during severe winds. Police would be mobilized and—this one was very important to Perez—volunteers would be deputized and authorized to arrest looters. Or to summarily shoot them if they chose to.

The plan's first test came on September 10, 1965, four years before Camille.

Hurricane Betsy, 1965

Initially, there was no reason for Luke or his fellow commissioners to pay much attention to the news on August 30 that a hurricane had developed out in the Atlantic Ocean and was heading toward coastal Maryland and Virginia. The next day the newly christened "Betsy" weirdly changed course and drifted west while building up strength, but by September 2 she was again moving northwest, this time aimed at the Carolinas. Meanwhile, the Weather Bureau and the navy, concerned that Betsy was growing both in physical size and in intensity, decided to experimentally seed her clouds to try to defuse the storm. Although most experts maintain that this seeding had no effect at all, a few contend that the law of unanticipated consequences may have reared its befuddled head during the experiment. In fact, what happened next was totally unexpected.

On September 5, when she was 300 miles due east of Jacksonville, Florida, and already north of the latitude of New Orleans, Betsy abruptly reversed her northwest motion and headed south. After ripping through the Bahamas, she skirted around southern Florida, and on September 8 she pummeled Key Largo with 120-mile-per-hour winds. Serious damage extended as far as the outskirts of Miami. Now heading west, Betsy entered the warm waters of the Gulf of Mexico.

It was time for Luke Petrovich to start paying attention, as he did any time a storm entered the Gulf. So far, the NHC and the Weather Bureau had gotten every aspect of their storm predictions wrong. They had forecast Betsy to make landfall the previous day in South Carolina—800 miles away, considerably to the north, and on the Atlantic coast. Yet, now the storm was entering the Gulf of Mexico, which rendered it certain to strike somewhere along the Gulf Coast. How much more wrong could the outside experts have been?

On the evening of September 8, the oil companies were already shutting down their rigs and bringing their men in. Luke phoned his fellow commissioners, got their agreement, then mobilized the police and road crews, the firemen, and the bus drivers. The Weather Bureau had yet to issue any official hurricane warning; all it had done so far was to declare a hurricane *watch* for a broad section of the Gulf Coast from Texas to Florida.

It was nonetheless clear to Luke, moving his cutouts around on his navigation charts, that either this was a very big hurricane or else it was moving mighty fast, or maybe both. The wind picked up. Shrimp boats heading north clogged the canals. Luke stayed awake through the night, coordinating the plan. The biggest task would begin at daybreak on September 9.

Best laid plans notwithstanding, it was a messy evacuation at best. The rains began shortly after dawn, much earlier than expected. Police slogged door to door in the mud, warning people to leave immediately. Some did, some refused, and some said they needed time to pack first. With wind speeds now around 135 miles per hour, Betsy was racing toward Plaquemines Parish at 17 to 20 miles per hour. The long evacuation caravan was not leaving the coastal region at anywhere near that speed.

The New Orleans radio and television stations issued the usual advice to stock up on batteries, candles, and canned foods and to fill bathtubs with water. By six o'clock that evening, it started raining in the city, even though the storm center was still out in the Gulf— nearly 50 miles south of the South Pass of the Mississippi River. The evacuees from southern Plaquemines Parish took refuge in public buildings in Belle Chasse, a town at the north end of the parish, or went on to New Orleans. Most folks in the middle part of the parish never did get to evacuate, because they were waiting hours for the caravan of refugees from the south to snail past. Meanwhile, Betsy had grown into a monstrous hurricane, packing sustained winds of 150 miles per hour, with an eye nearly 40 miles in diameter and winds of tropical-storm force extending out nearly 300 miles from her center.

Then, just prior to landfall, although the storm remained just as big and packed just as much moisture, Betsy's winds abruptly diminished to around 120 miles per hour. As the eye tracked up the west bank of the Mississippi River, some of the homes in Plaquemines Parish were partially sheltered from the eastern winds by the east levee. After the eye passed and the winds reversed, the west levee provided a slight windbreak. Nothing, however, could shelter anything south of the town of Empire from the nine-foot tidal surge that poured in from the Gulf. With most of the homes depending on no more than grav- ity to hold them in place, the floodwaters floated them off their blocks. Some would survive sufficiently intact to later be moved back to their original sites and repaired. Other were swept by the current into obstacles or into each other and smashed to smithereens.

At midnight, Betsy's eye passed just east of New Orleans. Numerous buildings in the city were damaged by hundred-mile-per-hour winds, but many more were damaged or destroyed by flooding. Parts of the city were submerged under as much as twelve feet of water, and several victims would be found drowned in the attics of one-story brick homes, which, of course, did not float. A major part of the destruction was due to the failure of the dikes along an industrial canal just east of the city, which allowed the backed-up Mississippi River to pour into the suburb of Chalmette. Shipping and port facilities along a hundred-mile stretch of the river were destroyed or badly damaged, and as far north as Baton Rouge a barge loaded with chlorine sank, which required officials to keep people away for a week before it could be recovered and secured.

At daybreak on September 10, Luke and the police found hundreds of survivors south of Port Sulphur huddled in and around the ruins of their homes with no drinking water, electricity, phone, food, or other basic needs and with dead animals and live fire ants and snakes all about. And so, on the afternoon after Betsy had passed, Luke took steps to complete the evacuation. This time, only men with official duties were allowed to stay behind.

Perez immediately set up roadblocks at the parish line and sealed off Plaquemines Parish from the outside world. No outsiders, whether newsmen or insurance adjusters, were allowed in. When Perez refused the governor's offer to provide National Guard troops, Luke attempted to reason with the old man. "Judge," he said, "we need those National Guard boys down here to give us a hand."

"A hand with what? We'll get the oil companies to help with the cleanup."

"That's fine," Luke said, "but first there's the issue of looting."

Perez twisted his bushy eyebrow. Thinking back to his liquor smuggling years, he well knew that there were a lot more ways in and out of the parish by water than by road and that there was no possible way to block all those waterways. "You're right," he relented. "I'll tell the governor to send the troops, but I'll insist that they be under civil authority." And so he did.

A few days later, the New Orleans *Times-Picayune* ran an internally contradictory story. In one column, it stated,

> Reporters [in Plaquemines Parish] saw no arrested looters. In most of the area there was nothing worth stealing.

Yet further in the same story was the following statement:

> "We've got the post office full of people arrested for trying to steal things," said Petrovich. "I just don't know how many. We haven't had time to count them. The Army patrols have been arresting them at night."

Luke, of course, was not about to publicly contradict the justification he had given Perez for needing the National Guard. This was simply an instance of the pragmatic Petrovich outmaneuvering the political Perez, with Luke realizing that the parish needed the National Guard for reasons other than preventing thefts. In fact, there was virtually no problem with looting. There seldom is after a major disaster. Besides, with Perez's reputation, what prospective looter would take the chance?

Full recovery from Betsy was slow and painful. New Orleans had gotten the lion's share of the publicity, and most of the aid and charity went there. The Army Corps of Engineers immediately took over the rebuilding of Chalmette's damaged industrial canal, but it was slow about doing anything in Plaquemines Parish other than cutting the back levee to drain the floodwaters. Meanwhile, the sixty-five hundred owners of damaged or destroyed homes and businesses got a rude surprise: none of their insurance policies covered floods. One after another, they argued with adjustors that their property had been destroyed by a hurricane—a windstorm—only to have their claims summarily denied and then denied again on appeal. Lucky were those few victims whose insurers prorated their claims, allowing 10 or 20 percent of the damages to be allocated to effects of the wind, with the remainder unrecoverable under the flood exclusions in their homeowners' policies.

When his constituents flocked to Perez for legal help, he organized a pool of attorneys to donate free legal assistance in settling their claims. He directed the parish council to allocate two hundred dollars per homeowner to move surviving structures back to their foundations. To restore the orange groves, he had the parish buy trees for citrus farmers on a matching basis. And he had the drainage system improved along the evacuation route.

But after 1966, locals saw less and less of "the Judge." The federal court injunction against interfering with school desegregation in the parish effectively silenced him locally. Then, in February 1967, he was devastated by the death of his wife. After making sure that his son Chalin would succeed him as council president, he announced that he would not seek reelection in that spring's primary. Remaining passionately committed to white supremacy, however, he spent the last two years of his life throwing his political and financial support behind the southern segregationist governors Lester Maddox and George Wallace.

Plaquemines Parish suffered two blows in 1969. In March, Leander Perez died of a heart attack; in August, Hurricane Camille devastated lower Plaquemines. The parish would face its first crisis in a half century without "the Judge" at the helm.

CHAPTER 5

STORM WARNINGS

On August 5, 1969, the National Hurricane Center in Miami received a stack of fuzzy weather satellite photographs from the national headquarters of its parent, the Weather Service in Washington, and they landed on the desk of Dr. Robert H. Simpson, the NHC's new director. A few years later, Simpson would collaborate with Herbert Saffir to create what is officially known as the "Saffir-Simpson Hurricane Potential Damage Scale," with its now familiar Category 1 through Category 5 hurricane ratings. But in 1969, there were no "categories" for hurricanes and the technology for monitoring the world's weather from space was still in its infancy. Nobody quite knew what to do with the images.

Simpson sifted through the photographs and noticed an inverted "V" cloud pattern off the coast of West Africa—a pattern that drifted westward during the next few days. Although this tropical wave initially showed no sign of circulation, Simpson alerted his staff to keep an eye on it.

If there was one sure benefit of this new satellite technology, it was that never again would a hurricane be "lost." That had always been embarrassingly difficult for meteorologists to explain to journalists and public officials, because it sounded as if the weathermen either had been grossly negligent or else had a short attention span. However, it was not uncommon for the first report of a tropical storm to scare all shipping out of a region, after which no further information would arrive because the storm no longer had any firsthand observers. Indeed, time and again, the consequences of such information gaps had been catastrophic. In the great Labor Day Hurricane of 1935, the storm was "found" again only after it was too late to evacuate the Florida Keys, and more than four hundred died there. Even more unnerving was the September 1938 hurricane that struck—of all places—New England, causing widespread devastation and killing about six hundred. In that case, there were no official warnings at all. That hurricane had been "lost" more than a day earlier, and in the interim it had taken an abrupt turn to the north and accelerated. And then, of course, there had been the devastating Hurricane Audrey in 1957, which, although never actually "lost," chose to misbehave egregiously during her final surge toward the coast after the last official update.

The first experimental weather satellite, Tiros-1, was launched on April 1, 1960, and it delighted its project scientists and engineers by returning 22,952 images before it malfunctioned seventy-two days later. The reliability of the instruments soon improved, and weather satellites launched after 1962 had lifetimes of two to four years. Soon every part of the planet was being photographed at least a couple of times every day. By 1969, eighteen weather satellites had been launched into orbit and five were still functioning. Camille would become the first hurricane they photographed continuously from birth through death.

Even those early satellites carried equipment that responded to infrared as well as visible light, so cloud tops could be photographed both day and night. The photographs were radioed to ground stations, where they were retransmitted to the Pentagon and the National Weather Service, which eventually passed them on to the NHC. Unfortunately, those images weren't very good by today's standards. Their resolution, low to begin with, was further corrupted by the analog scanning used to convert them to radio signals, yielding results that looked like they were copied from a slightly out of focus 1950s-era black-and-white television screen. An even more serious shortcoming was that the photographs lacked accurate reference points to identify the longitude and latitude, leaving this to be done manually at the receiving end. The result could be a large error (sometimes fifty miles or more) in pinpointing the center of a hurricane—information that was needed to calculate how far, how fast, and in what direction a storm moved during the time interval between a pair of photographs. The most serious shortcoming, however, was that the photos from space showed only the tops of the clouds, which were around eight miles above the vulnerable towns and cities. Although satellite imagery would lead to a deeper understanding of the nature of tropical storms as time went on, in 1969 it still didn't supply the kind of data the forecasters needed.

The previous year had been a good time for Bob Simpson to begin as director of the NHC. Although there were eight tropical storms that season and five of them made landfall, only one struck land with hurricane-force winds. That hurricane was Gladys, later (retrospectively) designated as a relatively weak Category 1, which caused minor damage as it crossed northern Florida from the Gulf to the Atlantic. The NHC under Simpson had issued adequate warnings, and no lives were lost. Those forecasts had been made in the conventional ways, combining data from ground stations with information supplied by reconnaissance aircraft. Although the satellite images hadn't contributed any actual data, they did play a role in providing Simpson with hints about the nature of the storm that informed his decisions to request those reconnaissance flights.

Now, the 1969 season's first tropical storm, Anna, had just been degraded to a tropical disturbance. If a second tropical depression continued to intensify, it would become tropical storm Blanch. Meanwhile, the still unnamed tropical wave from Africa continued to drift westward, and on August 9 it was about five hundred miles east of the Caribbean's leeward islands. On August 11 through 13, Blanch briefly escalated to hurricane intensity but then died out without striking land. Meanwhile, Simpson continued to pore over the images of the other westward-drifting tropical weather system, which seemed to be intensifying.

On Thursday, August 13, less than four days before Camille's landfall, the NHC's mother agency, the National Weather Service, wired the following forecast to newspapers from Louisiana to the Florida panhandle:

Extended Forecast

Louisiana, south half Mississippi, coastal Alabama, extreme northwest Florida— Temperatures Thursday through Monday will average near to three degrees below normal. Normal lows 70 to 75, normal highs 86 to 95. Mostly minor day to day changes. Precipitation moderate to heavy in mainly afternoon and evening thundershowers.

This was typical August weather that no reader would find unusual. Nobody, including the forecasters at the Washington office, had an inkling that a weather event was brewing that in just a few days would drastically alter the lives of tens of thousands of people, both in this region of the Gulf Coast and in a rural pocket of Virginia hundreds of miles inland.

On the morning of August 14, with the tropical disturbance now south of Cuba, Simpson asked the navy to dispatch one of its weather reconnaissance planes from Puerto Rico to investigate the system before it got out of their flying range. When that flight crew reported a central pressure of 29.50 inches of mercury and a wind speed of 55 to 60 miles per hour, Simpson upgraded the event to a tropical storm, assigned it the name Camille from the official 1969 list, and forwarded the preliminary assessments to Washington.

At noon (central daylight time, or CDT) on August 14, the NHC issued the first official advisory: Tropical storm Camille was located roughly 290 miles south of Havana, Cuba, and was moving west-northwest at 12 to 14 miles per hour with wind velocities of 50 to 60 miles per hour. The storm was likely to strike the western tip of Cuba. All interests in the Florida Keys and in the vicinity of the Yucatan Channel and the eastern Gulf of Mexico were advised to remain on alert. The dispatch added:

> While it is too early to determine what further land areas may be affected by this storm, the steering currents indicate the like-lihood of a turn to a slightly more northerly course Friday. This would carry Camille's center into the east central Gulf of Mexico this weekend. While Camille is a very small storm at present, conditions are favorable for further development.

For Robert Simpson, this was the beginning of, in his words, his "most critical and hair-raising experience" as director of the NHC.

Robert H. Simpson was not quite seven years old when the big hurricane of 1919 struck his hometown of Corpus Christi, Texas. When that erratic storm entered the Gulf of Mexico, the Weather Bureau immediately telegraphed storm warnings to all coastal stations from Texas to western Florida. People along those coasts soon noticed the growing surf. Then, for four days, the hurricane was lost. Exactly what it was doing during those four days was anybody's guess.

When Sunday morning, September 14, broke gusty with an occasional light rain, the Simpson family decided to forgo church services. By 9:45 a.m., the wind was screaming, and police began knocking on doors in the low sections of town—advising people to evacuate to high ground. Unfortunately, and despite the fact that there was hardly a person in town who hadn't heard about the terrible storm surge at Galveston nineteen years earlier, the general attitude in Corpus Christi that morning was that there was little danger and no reason to rush. After all, there had never been a major storm surge *here,* and the mainland was seemingly well protected by the long offshore ribbons of sand dunes known as Padre Island and San Jose Island.

As the Simpson family sat down to their Sunday dinner around noon, the sea surged into the low section of town and cut off the road to the North Beach region. The Simpsons watched in horror through their dining room window as a large house from across the street floated past, barely missed them, and crashed into the home next door, destroying both structures. Clyde Simpson scooped up young Robert as his brother-in-law steered Grandma Simpson's wheelchair into the flooded backyard. Robert's mother dumped the chicken and donuts from the dinner table into a paper bag and held it over her head as she waded into the rising water. The sight that was indelibly etched in young Robert Simpson's memory was that of his tired mother struggling to keep that bag of food above the waves, her arms growing ever weaker, until finally the bag met the water and the food floated away.

Nearly two hundred people drowned in the North Beach section alone; several hundred others died in the surrounding area. The Simpson family, its home situated close to high ground, survived. They returned after the storm to find their house filled with mud and debris, its walls smeared with tar and oil from the city's ruptured petroleum tanks. Young Robert saw death all around—dead dogs, cats, birds, and rats and the remains of one neighbor who hadn't made it out in time.

After that seminal experience, tropical storms were in Robert Simpson's blood. He studied at the University of Chicago with some of the world's leading atmospheric scientists, and then after college he joined the Weather Bureau, where he worked with the legendary meteorologist Grady Norton. Norton had an uncanny gift for forecasting that, Simpson noticed, relied on "personal intuitive skill, which few of his associates could replicate." During his own tenure as the first director of the Miami Hurricane Center, Norton himself admitted that

> Whenever I have a difficult challenge in deciding and planning where and when to issue hurricane warnings, I usually stroll out of the office onto the roof, put my foot on the parapet ledge, look out over the Everglades, and say a little prayer. By the time I return to the office, the uncertainties are swept away and I know exactly what my decision will be.

For Simpson, as much as he respected Norton the man, such a mystical approach would not do. After all, tropical storms are natural phenomena—every bit as natural as the motions of the planets—and, as such, they surely must be driven by natural rather than supernatural causes. To repeatedly "say a little prayer" and wait for God's guidance was expecting a bit much from a busy God.

Yet to scientifically predict the behavior of a hurricane, Simpson and the other forecasters needed to know a lot more about what goes on inside one. Bolstered by this view, in 1945 Simpson hitchhiked his first plane ride into a hurricane, and in the early 1950s he conducted a series of experiments from army weather aircraft in which he dropped instrument-laden balloons into storms. One of his first balloons, tracked by radar, survived for twenty-four hours as it was swept along by the winds.

Because Simpson's early hurricane research was not part of any formal program, it could claim no official priority for the use of military equipment. Meanwhile, the army's Air Weather Service needed positive publicity if it was going to be treated seriously at congressional budget hearings. Accordingly, in 1954 Major William C. Anderson, an enthusiastic public relations officer and himself a weather pilot, invited Edward R. Murrow of CBS to bring his film crew along on a flight into a hurricane. That storm was named Edna, and the flight was one from which Simpson had planned to launch instruments off

Cape Hatteras. Murrow accepted Anderson's invitation (over the objections of his network's producer), and Simpson's balloon experiment was preempted by the CBS film crew. "I was a bit teed off," Simpson would later admit.

From this flight, however, came dramatic film footage for Murrow's popular television show, *See It Now*. In that telecast, the celebrated journalist's voice resonated,

> The eye of a hurricane is an excellent place to reflect on the puniness of man and his works. If an adequate definition of humility is ever written, it's likely to be done in the eye of a hurricane.

Edna rampaged all the way into New England, pummeled Martha's Vineyard with winds of 120 miles per hour, crossed the coast near Cape Cod, and caused serious damage even into Maine. That disaster, coupled with Murrow's telecast, generated a windfall of nationwide interest: youngsters wanting to become meteorologists, veteran World War II pilots seeking jobs in weather reconnaissance, and the public writing congressmen to support funding for hurricane research.

Senator Theodore Green (D-RI), incensed that a hurricane could strike his constituency so far north without anyone understanding *why*, publicly criticized the Weather Bureau for not developing a comprehensive program for hurricane research. The Weather Bureau's chief, Francis W. Reichelderfer, responded that he had indeed proposed precisely such a research program in his budget for three years in a row but each time the Eisenhower administration had deleted it. When the senator asked who at the Weather Bureau had actually developed that research proposal, Reichelderfer swallowed. It was a mid-level meteorologist named Bob Simpson, a fellow who had been conducting bootlegged experiments from army recon flights for the past eight years, without any formal appropriations. Green asked to meet the man.

At their meeting, Simpson made an enthusiastic and compelling case for the importance of meteorological research on hurricanes and their prediction. Senator Green had already been convinced, but now he was also impressed. He drew together a group of legislators from other states that had been affected by the 1954 hurricanes and scheduled hearings on the hurricane problem. The principal witnesses were Francis Reichelderfer and Harry Wexler, the Weather Bureau's director of research, but Simpson was also subpoenaed as a technical expert. Although Reichelderfer and Wexler were politically handcuffed into supporting the Bureau of the Budget at the hearings, Simpson was under no such obligation, and he made an eloquent summary of the research plan the Eisenhower administration had summarily rejected. Based on those hearings and Simpson's personal initiatives, Congress created the National Hurricane Research Project (NHRP). The Weather Bureau's budget, $27.5 million in 1954, jumped to $57.5 million in 1955.

Simpson continued his hurricane investigations for the next thirteen years. When there wasn't an actual hurricane around to study, he'd comb through the archives to try to find patterns of similarity in the behaviors of historical storms. With the advent of computing, he worked with programmers to develop mathematical models of hurricane formation.

Above all, what drove Robert Simpson was a worldview that, at the deepest level, Mother Nature is not capricious. Hurricanes are natural events, and natural events have natural causes. Investigate deeply enough and that web of causes, regardless of how complex, will ultimately yield to the mind of men and women. Yes, women too, for Simpson's own wife, Joanne, was a Ph.D. meteorologist specializing in tropical storms.

Now, in August 1969, some thirteen years older and entering middle age, Robert H. Simpson found himself directing the NHC. A bit had been learned in the interim, but scientifically the going had been slow. Limited funds for research notwithstanding, hurricanes simply didn't eagerly give up their secrets. They're not the kind of phenomenon you can re-create in the laboratory and study over and again, refining your measurements each time and publishing your observations and analyses that send other laboratories around the world scurrying to replicate those findings in their own labs. Yet Simpson maintained his steady faith that there are underlying physical principles that drive all large-scale meteorological events, including—and particularly—hurricanes.

The contemporary developments in computing offered promise,

but barely more than promise, because computers need programs, and programmers need mathematical models, and such models are developed by analytical scientists, and those scientists need data, and data requires instrumentation, and instrumentation design is a matter of engineering, and engineers must know what the instruments they design are supposed to do—and so the whole picture wasn't very complete to anyone. As for those multimillion-dollar satellite photos, Simpson examined them again, but they still didn't answer the most important questions: *where* will this hurricane strike land and with what *effects*? The only thing the photos revealed was where Camille was at the time each shot was taken and what she looked like from a vantage point several hundred miles above sea level.

Yet Simpson and his staff knew they couldn't be the least bit neglectful. If the storm continued to curve as it was doing, it would pass through Havana and ultimately slam into northwestern Florida. At 2:00 p.m. on Thursday, August 14, two hours after the initial advisory, Simpson sent off a precautionary bulletin for western Cuba. At 5:00 a.m. on August 15, with the winds hovering around 70 miles per hour, he repeated the same precautions. Then, in the next couple of hours, the winds escalated to 90 miles per hour, and Camille was a small but potent hurricane. By early afternoon, central winds were up to 115 miles per hour and gale-force winds extended 125 miles north of the eye and 50 miles to the south. Camille struck the western tip of Cuba at about 6:00 p.m. on August 15, dumping ten inches of rain on the region while the frictional drag of that landmass weakened her winds to 92 miles per hour.

At 11:00 p.m., the ground radar installation at Key West reported that Camille had entered the Gulf of Mexico just north of western Cuba. Radar itself, however, cannot locate a hurricane's eye; all it can do is detect where the rain is falling. Although it's usually the case that a nonreflecting region surrounded by circular bands of rain is indeed the eye of a hurricane, the interpretation isn't always so direct. Hurricanes sometimes have very little rain in their southwest quadrant, and in these instances the storm echo does not reveal the center.

Moreover, the eye of a hurricane is a dynamic phenomenon—it can grow, shrink, distort its shape, and stagger around like a drunken sailor. A single radar image can give a misleading impression of where a hurricane is centered, and a single pair of such images can give a wildly misleading view of where that center is headed.

Simpson, of course, knew all this, as did his staff meteorologists who pored over the radar images. But within an error of perhaps 50 miles, he figured it was safe to say that Camille lay about 250 miles south-southwest of Key West and 700 miles southeast of New Orleans. The storm was certain to strike somewhere along the Gulf Coast within a few days, and by then it could easily experience a growth spurt.

At the time a tropical storm acquires a name—when its sustained winds exceed 39 miles per hour—the NHC begins issuing advisories at no less than six-hour intervals, with special bulletins at shorter intervals as appropriate. As of Friday evening, Simpson had already issued six advisories and seven bulletins about Camille. In Advisory No. 7, issued at 11:00 p.m. CDT on Friday, August 15, he inserted the statement, "A hurricane watch will probably be issued for a portion of the coastal area of the northeast Gulf by or prior to noon Saturday." Meteorologists at the NHC unfolded cots and unrolled sleeping bags to bunk in their offices for the next few nights, where they would be ready on a moment's notice to analyze any new developments.

At 8:00 a.m. CDT on Saturday, August 16, Simpson issued a hurricane watch for the Gulf Coast from Biloxi to St. Marks, Florida. Several newspapers that morning elevated Camille to front-page status. Newspaper deadlines being what they are, however, the published reports ran considerably behind the unfolding event. The New Orleans *Times-Picayune,* for instance, included a tracking chart that showed Camille's glancing blow on Cuba, accompanied by a single-column story:

Hurricane Churns into Gulf Waters

Danger to Florida Still Not Determined

Hurricane Camille, the most intense tropical storm since Beulah ravaged the lower Rio Grande Valley of Texas two years ago, battered the extreme western coast of Cuba with 115 mile-an-hour winds Friday night.

> Dr. Robert Simpson, head of the National Hurricane Center at Miami, said it will be 36 to 48 hours before it can be determined whether Camille will be a major threat to the Florida mainland.

By the time any readers read this story, the prospective hurricane watch had long been issued.

But a watch is one thing; a *warning* is another. A watch is an alert stating that a coastline *may* be struck by a hurricane within the next thirty-six hours. A warning implies that sustained winds of at least seventy-four miles per hour are imminent on a coastline within about twenty-four hours. Although the NHC has no authority to order evacuations, it is generally assumed that a hurricane *warning* will mobilize local and regional authorities into directing their own residents away from imminent danger.

An evacuation, however, is a disruption whose economic impact can easily run to a million dollars a mile along a populated coast. Who should be warned, and exactly of what should they be warned? To make such decisions, local officials needed fairly specific information about the risk. Simpson, in turn, needed aerial reconnaissance to report on the storm's intensity, its size, and its central pressure, before he could make any predictions that would be better than educated guesses.

Unfortunately, no reconnaissance aircraft were available. The navy, which at that time was responsible for probing hurricanes in the Gulf, had ordered most of its weather planes and crews to Puerto Rico to fly into another storm in the Atlantic. That mission was part of the government-sponsored Project Stormfury, an expensive experiment to evaluate the prospects of weakening a hurricane by seeding its clouds with silver iodide. Considerations of scientific methodology had led to the decision to seed only those storms that had no chance of striking land within the next few days. The methodological logic was this: Storms *always* fizzle out when they hit land whether they've been seeded or not, so to evaluate the effects of seeding, the scientists needed to target storms that had little chance of reaching land. Thus, to his horror, Simpson was informed that all of the instrumented planes with trained crews were buzzing around out in the Atlantic, a long way from where they were needed to probe Camille.

He pleaded with the navy to do whatever they could. Word came back that there were two older instrumented planes left in Jacksonville, and as soon as qualified crews could be pulled together they would be dispatched into Camille. One of those planes eventually did get to the hurricane, but its crew reported that the storm was too strong to penetrate and returned. The other plane never got off the ground.

Meanwhile, more satellite photos arrived. By today's standards their resolution was crude, and when laid over a map to get the storm's coordinates at a specific time, the error could easily amount to fifty miles in any direction. Moreover, as with every other satellite photograph since, they showed mainly the tops of the clouds, which were about forty thousand feet above the sea. Today there are mathematical models for predicting sea-level winds from satellite images, but in 1969 the whole concept of understanding (let alone predicting) weather by satellite was in its infancy. As Simpson would later reflect,

> There I was, trying to make up my mind on how to put up the warnings for this thing. The people in Washington who were analyzing the satellite images were sure the storm was losing intensity. I was sure the storm was getting stronger. I was convinced it was becoming close to a record storm just from the way the eye structure was changing. We could see the structure, but not what the central pressure was, or what the strongest winds were.

With little more to go on than his intuition, Simpson issued a hurricane warning for the Florida panhandle from Fort Walton to St. Marks (just south of Tallahassee), with a gale warning elsewhere from Pensacola to Cedar Key. Coastal residents should prepare for tides in the range of five to ten feet. That warning, part of Advisory No. 10, reached the radio and television stations around noon on Saturday, August 16.

By then, the skies of the Florida panhandle were already filled with aircraft being evacuated from Eglin and Tyndall Air Force Bases and from the Pensacola Naval Air Station. Several marine contractors in the region took the expensive precaution of intentionally sinking their barges to prevent them from damaging shoreline structures in a storm surge, planning to refloat them after the storm passed. Panhandle residents clogged the phone lines, quickly scooping up every motel room within a hundred miles inland. Lumberyards sold out all of their plywood, and supermarkets ran out of batteries, candles, and toilet paper—the latter reflecting a curious mass behavior that always seems to accompany a hurricane warning. Most small craft owners pulled out their boats or motored them up canals and rivers. Some unluckier ones sailed their vessels to ports to the west. The best estimate was that Camille would arrive in about thirty-six hours, but it could be as few as twenty-four. Some residents left immediately to beat the crowd. Most decided to wait until Sunday morning.

The NHC desperately needed reports from airplanes—strong airplanes—with accurate instrumentation and experienced and fearless crews who would competently compute and radio back the essential storm information. Unfortunately, nobody had thought to create a system where such a resource would automatically be at Simpson's disposal in a crisis situation. Simpson got on the phone with the commander of the Air Weather Service, located at Scott Air Force Base in Illinois, and described what was at stake, not only for the military but for millions of civilians along the Gulf Coast. The commander was sympathetic, cooperative, and honest. "I think I can help you," he said, "but it's going to take eight hours before I can get something to you."

Meanwhile, Camille steadfastly refused to cooperate with the predictions. Between about 1:00 p.m. and 8:00 p.m. that Saturday, with winds up to an astonishing 150 miles per hour, she essentially quit moving, wiggling around a bit but showing no inclination to veer toward Florida. When a tropical cyclone stalls like this, it means that competing forces are struggling to steer it in different directions. It's a meteorologist's nightmare, and there is little one can do but pay close attention until something more happens.

In Advisory No. 11, issued at 5:00 p.m. CDT on Saturday, August 16, the NHC reported that Camille was stalled about 380 miles south of Fort Walton, Florida, with winds of 150 miles per hour. In the same advisory, Simpson raised the storm surge estimate slightly, to the range of five to twelve feet. Most bets remained that, when the hurricane did start moving again, it would veer away from Louisiana and Mississippi and toward the Florida panhandle. It was along that coast that flooding was predicted.

That evening, one of the U.S. Air Force's C-130s from Illinois reached the storm, penetrated it, and reported a central pressure of 26.72 inches of mercury. This was flirting with the U.S. record low of 26.35 inches, set in the Florida Keys Labor Day hurricane of 1935. At Camille's latitude, however, the air is slightly heavier than at the latitude of the Keys, so in effect Camille's intensity was a bit greater than that of the Labor Day event. And, unlike that earlier terrible storm, which swept through a few sparsely populated islands, Camille was certain to strike the mainland.

As for the wind speed, the C-130's flight engineers reported an astounding 160 miles per hour, sustained, with higher gusts. In its midnight advisory, the NHC reported that Camille was an "intense" hurricane centered about 310 miles due south of Pensacola, Florida. And now the storm was again on the move.

Simpson and his staff computed and cogitated into the night, combining the sketchy prior information with what they'd learned more recently from the reconnaissance flight crew. Camille was traveling north at ten to twelve miles per hour while drifting westward at six to seven miles per hour. If her movement continued at the same rate in the same direction, that straight line would pass just south of New Orleans and the highest winds and worst surges would hit the Crescent City in twenty-three to twenty-seven hours.

But the tropical storm specialists also knew that every hurricane heading north is deflected to the east by the Coriolis force, a natural consequence of the Earth's rotation. Although upper-level winds, so-called steering currents, can sometimes retard this general eastward curvature, all of the current data from land stations suggested that the existing steering currents would assist rather than hinder Camille's eastward deflection. A big unknown was the sea temperature, how deep it ran, and whether Camille might track over macro-vortices of warm water that were undetectable with 1969 technology. The NHC computer programs could do no better than the data they were fed.

Even under the best circumstances, those early computer forecasts were not particularly accurate. In 1969, the average error for twenty-four-hour predictions of a hurricane's movement was 129 nautical miles (148 statute miles). Roughly half of this error was attributed to inaccuracies in measurement. The center of a hurricane could not be positioned more precisely than about plus or minus 27 nautical miles, and this alone could generate an error in the direction of movement on the order of ten degrees or so. On top of this lay the whole issue of varying environmental conditions: those steering currents and seawater temperatures that certainly were affecting Camille's development and behavior but that the computer models of the time could account for only coarsely. Computers were certainly a benefit, but their answers needed to be critically evaluated in terms of all of the uncertainties.

Simpson mulled over the evidence and concluded, again, that in all likelihood Camille would strike somewhere along the Florida panhandle shortly after nightfall on August 17 and that points west would experience the offshore winds of the milder left side of the vortex. The storm surge would be mainly to the east.

Meanwhile, Chester Jelesnianski, a meteorologist who had been hired as part of the National Hurricane Research Project funded by Congress, came to Simpson's office with a sheaf of green-bar computer printouts. He'd been working on the problem of predicting storm surge heights and had developed a mathematical model that for the first time took into account the effects of the coastal sea bottom. The model had yet to be validated through actual observations. "I thought you might be interested in these," he said.

Simpson studied the figures. If Camille were to strike west of Biloxi, near Pass Christian, the computations predicted a storm surge of twenty-five feet above normal sea level. "This is half again as high as anything that's ever been recorded," Simpson said.

Jelesnianski diplomatically corrected him. Surges of twenty-five feet, and even more, had been documented in East Pakistan (now Bangladesh) and India, where tropical cyclones often sweep north into the Bay of Bengal. Hundreds of thousands of people had died in some of those catastrophes. Given that the bottom slope and coastline of the Mississippi Sound bears some resemblance to the site of those Asian disasters, a surge of twenty-five feet on the Mississippi coast during a strong hurricane was not beyond the realm of possibility.

"Interesting," Simpson said, laying down the printouts. "But Camille isn't going to hit west of Biloxi."

"Just in case it does, Bob, I thought you'd like to know what the model is predicting."

The seemingly simple concept of "sea level" becomes a lot less simple when you try to define it precisely. Stand at a wharf and watch the water height on one of the pilings. In the short term, wave crests and troughs ripple the surface. Then, over the course of a day, the average level rises and falls twice—a consequence of the gravitational interaction between the sea and the moon. These tides, however, are not the same from day to day or month to month; some are higher and some are lower, depending on the relative positions of the sun and the moon and the inclination of the Earth's axis to the direction of its orbital motion around the sun. And then the weather also has an effect. Sometimes a big one.

The idea of sea level is based on an idealized calm sea—one with no waves or tides, with no wind blowing over it, and with the atmosphere pressing down with a "standard" pressure of 14.69595 pounds per square inch, which is equivalent to the pressure of a column of liquid mercury 29.92126 inches in height at sixty-eight degrees Fahrenheit. Of course, a real sea never conforms to this exact combination of conditions. Thus, except for a fleeting instant every now and then, the sea we see is never precisely at "sea level."

But just as we find the concept of a straight line useful despite the real-life impossibility of such a geometrical abstraction, so does the concept of sea level have its usefulness. By knowing where the sea *ought* to be, we can separately describe the deviational effects of waves, tides, atmosphere, and storm surges. Such information allows engineers to make informed decisions about the design and placement of shoreline structures from wharves to roads, homes to businesses. It enables emergency planners to define and prepare for worst-case storm scenarios. And, over the long term—decades or more—such information even informs scientists about the conjectured effects of climate changes such as global warming.

Every introductory physics textbook offers elegantly simple mathematical descriptions of wave phenomena—descriptions that apply very nicely to sound waves and light and radio waves, all easily verified in the laboratory. Yet seldom do such books delve into water waves, which are much more obvious to the human senses. And for good reason: water waves are complicated. They behave differently in deep water than in shallow water, they "break" on shores in ways that textbook waves do not, and they create surf and run-up currents that defy mathematical description. They also impact humanity in profoundly different ways than the mathematically ideal waves described in most textbooks.

Water waves arise when the surface of the sea is disturbed, whether by an earth tremor, an undersea landslide, a passing ship, or the wind. As the sea bobs up and down, it transmits the local disturbance to the adjacent water, which is also set to rising and falling, which then sets the water further out bobbing up and down, and so on. In deep seas, the water itself moves in a vertical circle while the energy is carried horizontally in the pattern we recognize as a wave. Usually what we observe, however, is not a single set of crests and troughs but rather a complex superposition of many waves from many sources, all traveling in different directions with different heights and speeds. The sea's often chaotic surface is the sum of the momentary heights of all the waves that are passing through a given point at a given instant. And that pattern is continuously in flux.

Blow across a loose piece of paper held in front of your lips, and the paper will rise into the airstream. Set a wind blowing across a sea, and the water will also rise in a hump—a hump that initiates a wave. The longer the wind blows, the higher the wind speed; the greater the "fetch" (the distance it blows over), the taller the waves grow.

In deep water, a wave must be extremely energetic to "break." And when this does happen, the wave usually breaks backward; that is, an explosion of foam and surf tumbles down its backside. Thus, a ship at sea can usually survive even the worst of storms, provided that the wind and waves are traveling in the same direction (they usually do, more or less) and provided that the vessel can keep its bow pointed into the waves. The ship then rides up the oncoming swell, over the top, and down the backside in the same direction as any breaking water, plunges into the trough (where it may have its decks washed over as it levels out), and then rides up the next swell.

In shallow water, however, the dynamics change. When the water depth is less than about ten wavelengths (the horizontal distance between crests), the motion of the wave extends all the way to the sea floor. Now the wave drags against the bottom, its forward speed decreases, and it grows in height. It also bends (or "refracts") until its crests are roughly parallel to the shoreline, regardless of the direction of the wind. All of the separate waves of different wavelengths and wave speeds that contribute to the complexity of the surface of the deep sea now fall into a lockstep pattern, carrying their energy toward shore with one single speed. And then, when the water shallows out to where it can no longer sustain the wave's height, the wave rolls forward and breaks into an explosion of surf.

It is during its death throes, of course, that a wave does its damage. A water wave carries a great deal of kinetic energy, and when such a wave disappears at the surf line, that energy must go somewhere. Some of it is converted into heat, a little bit goes into sound, and whatever is left goes into dislocating heavy objects it encounters— beach sand, riprap, piers, wharfs, buildings, roads, bridges, ships.

We humans have a habit of thinking linearly: we go to a beach, we watch what a three-foot breaker does, and most of us assume that a six-foot breaker would do twice as much. Mother Nature, however, is not so simple minded. Each time she doubles the height of a wave (all other factors being equal), she packs four times as much energy into it. If we compare a twelve-foot wave with a three-foot wave, the larger one carries roughly sixteen times as much destructive power. Large waves striking a shoreline are extremely potent agents of potential destruction.

But this is still not the whole story about waves. In a hurricane, something additional happens to the sea: the phenomenon of storm surge.

A storm surge begins when a broad bulge of seawater, typically around fifty miles in diameter and a few feet high, is lifted up by hurricane-force winds. This bulge follows along beneath the storm, its height somewhat greater in the storm's right front quadrant, where the winds are the greatest. When the storm enters a coastal shallows, the bulge drags against the seafloor, slows its forward motion, and the faster-moving tail end of the swell piles up and over the front end, increasing its depth. Meanwhile, the normal storm waves still ply the surface.

If hurricanes are monstrously complicated to predict, storm surges are no easier. The height of a surge depends on complex interactions between the atmosphere, the sea, the seafloor and its slope, and the specific geometry of the coast. The worst possible combination of conditions is an intense but compact hurricane sweeping into a shallow-water bay—exactly the set of conditions that Bay St. Louis in

Mississippi offered to Hurricane Camille and that Chester Jelesnian-ski had warned Bob Simpson about.

Unfortunately, until almost the last minute, nobody figured that Camille was destined to strike Bay St. Louis of all places. The official advisories and bulletins had assured everyone that the storm would strike considerably to the east. The last advisory on the night prior to the disaster, less than twenty-four hours before landfall, read in its entirety:

ADVISORY NO. 12. 11 PM CDT SATURDAY AUGUST 16, 1969

. . . CAMILLE . . . EXTREMELY DANGEROUS . . .
THREATENS THE NORTHWEST FLORIDA
COAST . . .

HURRICANE WARNINGS ARE IN EFFECT ON THE
NORTHWEST FLORIDA COAST FROM FORT WALTON
TO ST. MARKS AND GALE WARNINGS ELSEWHERE
FROM PENSACOLA TO CEDAR KEY. PREPARATIONS
AGAINST THIS DANGEROUS HURRICANE SHOULD BE
COMPLETED SUNDAY MORNING. A HURRICANE
WATCH IS IN EFFECT WEST OF FORT WALTON TO
BILOXI.

WINDS WILL INCREASE AND TIDES WILL START TO
RISE ALONG THE NORTHEASTERN GULF COAST SUN-
DAY. GALES SHOULD BEGIN IN THE WARNING AREA
SUNDAY AND REACH HURRICANE FORCE IN THE
FORT WALTON ST. MARKS AREA SUNDAY AFTERNOON
OR SUNDAY NIGHT. TIDES UP TO 15 FEET ARE
EXPECTED IN THE AREA WHERE THE CENTER
CROSSES THE COAST. TIDES ARE INDICATED 5 TO 12
FEET ELSEWHERE IN THE HURRICANE WARNING
AREA.

ALL INTERESTS ALONG THE NORTHEASTERN GULF
COAST ARE URGED TO LISTEN FOR LATER RELEASES.

AT 11 PM CDT . . . 0400Z . . . HURRICANE CAMILLE
WAS LOCATED NEAR LATITUDE 25.8 NORTH . . . LON-

GITUDE 87.4 WEST . . . OR ABOUT 325 MILES SOUTH
OF PENSACOLA FLORIDA. CAMILLE WAS MOVING
NORTH NORTHWESTWARD ABOUT 12 MPH. A
CHANGE TO A MORE NORTHERLY COURSE IS INDI-
CATED WITH LITTLE CHANGE IN FORWARD SPEED.

HIGHEST WINDS ARE ESTIMATED 160 MPH NEAR THE
CENTER. HURRICANE FORCE WINDS EXTEND OUT-
WARD 50 MILES AND GALES EXTEND OUTWARD 150
MILES FROM THE CENTER. CAMILLE IS EXPECTED TO
CHANGE LITTLE IN INTENSITY DURING THE NEXT 12
HOURS.

SMALL CRAFT FROM PENSACOLA TO CEDAR KEY
SHOULD SEEK HARBOR . . . AND SMALL CRAFT ON
THE ALABAMA . . . MISSISSIPPI AND SOUTHEAST
LOUISIANA COAST SHOULD NOT VENTURE FAR FROM
SHORE.

CHAPTER 6

ON THE COAST

Plaquemines Parish, August 16, 1969

That Saturday morning was a good day for fishing, and Luke Petrovich was offshore in his boat when he noticed an unusually large number of helicopters shuttling to and from the oil platforms. A fellow parish commissioner raised him on the radio. The latest news: a hurricane watch had been declared from Biloxi to the Florida panhandle.

Luke unrolled his master chart of the Gulf. "Do you have the coordinates?"

"Twenty-four five, eighty-six zero, headed north-northwest at ten."

Luke studied the map. "That sends it straight here."

"The feds say it's gonna turn north. Do you believe 'em?"

He glanced up at the helicopters. "Nope."

It was the kind of thing that could make a rational man's blood boil: a hurricane headed for southeastern Louisiana yet no official warning from the National Weather Service. Not that Luke was surprised by this kind of neglect from the feds. As Leander Perez had repeatedly pointed out, Washington bureaucrats got concerned about Louisiana only when they wanted to meddle in issues like school desegregation or voting rights for coloreds. Expect those bastards to do something *useful* for southern folk and you'll be waiting 'til kingdom come. The lack of timely hurricane warnings was just another example.

It took an official vote of three of the five commissioners to mandate an evacuation, and Luke scheduled the necessary meeting by ship-to-shore radio. At Port Eads, he briefly tied up and dashed into the pilot house to phone the New Orleans office of the Weather Service, asking what was up. Whatever transpired during that conversation, Luke would later decline to share any details. Despite his anger at the time, he apparently felt a need to be protective of whomever he talked to. Later, he was more sanguine. After all, that conversation turned out to be irrelevant, so why try to dredge it up more than three decades later?

Luke had already learned to take his cues from the oil companies, bolstered by those petrol boys' excellent track record for knowing what they were doing when weather threatened their rigs. He learned that they'd started tapering off production on Thursday, that everything was completely shut down by the time he reached shore, and that all of those helicopters were bringing in more than three thousand offshore workers. The rigs would be fully evacuated by Saturday afternoon—a monumental task involving hundreds of flights.

Luke motored upriver to Pointe A La Hache and scampered over the levee to the courthouse. The decision was made with no debate. By that afternoon, deputies were cruising the parish neighborhoods, their loudspeakers blaring evacuation preparation announcements. Yet another ten hours would pass before the National Hurricane Center would declare even a hurricane watch (let alone a hurricane *warning*) for the region.

Later, in its official report on the aftermath of the disaster, the U.S. Army Corps of Engineers would describe how effective this evacuation had been, with 17,800 people successfully leaving the birdsfoot delta in a timely manner in a slow-moving caravan on the single narrow two-lane road to the north. The official federal report would fail to mention, however, that this mass departure, which certainly saved thousands of lives, took place with little thanks due to the advisories from the NHC. Had the locals waited for the feds' official hurricane warning, they would have had only four hours or less to drive as far as seventy-five miles on a single clogged road, and hundreds or even thousands would have drowned in their cars when the storm-driven Mississippi River backed up and overflowed its west levee.

The Mississippi Gulf Coast

Residents of coastal Mississippi didn't have the history of evacuation experience that the folks near the mouth of the Mississippi River did, where getting out entailed a long drive north on a single road vulnerable to flooding. In Mississippi, the coast ran east and west for seventy-five miles and spanned three counties. The main road, U.S. Highway 90, paralleled the beach from Pascagoula west through Ocean Springs, Biloxi, Gulfport, Long Beach, Pass Christian, Bay St. Louis, and Waveland. But there was also a handful of roads heading north, and coastal evacuees at least had a choice of more than one way to get inland. It was also possible to get onto Interstate 10 near Gulfport and head east, although to the west that highway was still incomplete at that time.

In 1969, life on the coast had a leisurely pace. Commanding grand views of the Mississippi Sound from the north side of Highway 90 stood hundreds of stately homes, many of them antebellum, a large number serving as summer residences for wealthy New Orleanians. Although ocean freighters—mainly banana boats—did dock at the deepwater facility in Gulfport, most of the commercial marine traffic consisted of shrimp trawlers. Dozens of churches, including several black churches, stood on property that today would be beyond the financial means of any congregation to purchase. The Gulfside Assembly in Waveland, a black Methodist retreat center, sprawled over seventy acres along the Gulf and served as one of the few places in the state where black civil rights activists could meet with relatively little harassment. Except for a few low-lying neighborhoods along the back bay in Biloxi, however, it was the whites who lived closest to the water, the blacks generally living a few blocks or more inland.

The most striking feature of the region was its huge beach: 100 to 150 yards deep, stretching virtually uninterrupted on the south side of Highway 90 for the twenty-six miles between Biloxi and Pass Christian. The few structures on this vast expanse of sand included a couple of docks and marinas and the port facilities at Gulfport.

As natural as the beach looked (and still looks), however, it was artificial. In its natural state, this shoreline consisted of muddy tidal flats that would sometimes be inches below the water, sometimes inches above, and that would be reconfigured with every high spring

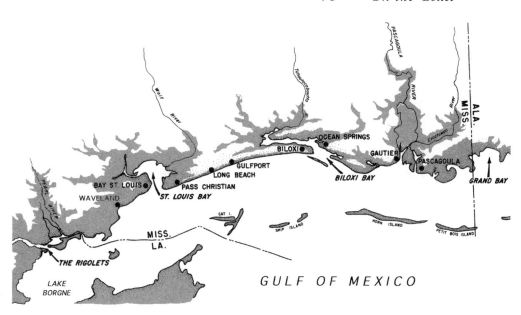

The Mississippi Gulf Coast. *(Adapted from USACE, Mobile District.)*

tide or storm. In 1924, construction commenced on a concrete sea-wall to prevent the recurring erosion of Highway 90. When the hurricane of 1947 demonstrated the inadequacy of that tactic alone, the current twenty-six-mile beach was created as a buffer zone against the waters of the Gulf. This huge project was financed through a combination of a local bond issue (paid for through a two-cent-per-gallon gasoline tax) and funding from the federal government. Because of the federal money involved, the entire beach had to be designated for public use in perpetuity. And so it was. Everyone could use all twenty-six miles of the newly created beach, as long as they were white. As for colored folk, well, they could use a few dozen yards of beach in front of the Gulfport Veterans Hospital.

The economy wasn't exactly booming in 1969, but the cost of living was low. Military personnel stationed at Keesler Air Force Base and the U.S. Naval Construction Battalion Center pumped money into local businesses, there was some work to be had at the wharves or with the shrimpers and packers, and there were numerous domestic and grounds-keeping jobs at the estates. Compared to the rest of the state of Mississippi—with the lowest per capita income in the nation—the Gulf Coast wasn't a bad place to live, at least for whites.

There was also something new in the local economy. Just eight years earlier, when President John F. Kennedy announced his ambitious plan to put Americans on the moon before the end of the decade, NASA approved a 13,500-acre site surrounded by a sound buffer of 125,000 acres of marshland for a place to conduct static tests on the launch vehicles. The Mississippi Test Facility (later renamed the John C. Stennis Space Center) was the largest construction project in the state of Mississippi and the second largest in the nation at that time. It was here that the Apollo program's Saturn V rocket engines were first fired and flight certified. On July 20, 1969, just a month before Camille, the Apollo 11 mission successfully carried a crew of three astronauts to the moon, two of whom walked on the lunar surface.

For the Gulf Coast businessmen and bankers, the arrival of the space center was the best news imaginable. They immediately began circulating the slogan "If you want to go to the moon, you first have to go through Hancock County, Mississippi!" The imminent tidal wave of tens of thousands of engineers and scientists (which was wildly optimistic by any standard) would be a boon to the local economy, the reasoning went. Other locals—those who valued the simpler life— were not so sure.

Those who would promote the Gulf Coast, however, were quite aware that they had an image problem. Statewide, for instance, Mississippi's per capita income in 1969 was only 62.1 percent of the national average. And of the fifty states, Mississippi ranked at or near the bottom on every statistic most outsiders would relate to quality of life: education, transportation, hospitals, life expectancy, cultural outlets, and so on. To top this off, there was also a considerable amount of bad publicity about race relations.

Much, if not most, of the state's negative reputation for racial conflict was based on fact. Racial tensions began to escalate after Mississippi blacks, along with millions of other young males across the nation, were discharged from the armed services after World War II. After being welcomed as equals and even heroes by the European allies, black veterans returned home to find themselves segregated, mistreated, and demeaned by white Mississippians. They now knew better, however, than to simply shrug and accept an inferior social status as their proper place in life.

The philosophy of white supremacy wasn't monopolized by common folk; it extended to the highest levels of Mississippi government. In 1944, for instance, Mississippi's U.S. senator Theodore Bilbo introduced a bill in Congress that would deport all African Americans to Africa— regardless of whether they had performed with distinction in the U.S. armed forces. Two years later, Bilbo explicitly promoted violence against blacks, stating publicly, "Remember the best way to keep the niggers from voting. You do it the night before the election." In 1947, officials in Jackson and Hattiesburg forbade a scheduled stop of the Freedom Train, because that rolling showcase of democracy sponsored by the federal government had no provision to prevent unsegregated viewings of documents such as the Declaration of Independence. The list of examples of such government-endorsed bigotry goes on and on.

In 1960, northern civil rights activists shrieked in disbelief when Mississippi voters elected Ross Barnett as governor. A storekeeper whose only qualification was his bombastic yet bumbling rhetoric supporting "segregation forever," Barnett entered the office with an undistinguished high school education and served with a talent for making himself look foolish to the press. Under Barnett's watch, the federal government sent troops to the University of Mississippi to enforce a court order admitting a black student, James Meredith. Barnett's outrageous public statements contributed to the ensuing riots on that campus—the state's flagship university—which did irreparable harm to the national reputation of "Ole Miss." When the Klansman Byron De La Beckwith was tried for the 1963 murder of Medgar Evers, an NAACP officer, Governor Barnett showed up at the trial and shook hands with the murderer in front of the jury; the case resulted in two hung juries before Beckwith was finally convicted of the murder some twenty-five years later. During the Mississippi Summer Project of 1964, thousands of Yankee college students invaded the state to help blacks register to vote. Barnett did nothing to quell the resulting violence, and his rhetoric actually condoned much of it—which included the arrest of more than a thousand activists; some eighty beatings; thirty-five shooting incidents involving blacks and civil rights workers; three nationally publicized murders in Philadelphia, Mississippi; thirty-five black church burnings; and thirty bombings of homes and offices. During Barnett's single term, he accomplished nothing positive other than to catapult his name into national prominence, where it became synonymous with bigotry and stupidity.

Some apologists in Mississippi shrugged that the activists need not have alarmed themselves at Ross Barnett's election and that the man's big mouth actually had little to do with the violence. Although the governors of other southern states, particularly Louisiana and Alabama, were always powerful figures, in Mississippi the governor was essentially an irrelevant ornament. The Mississippi governor had only a meager handful of executive powers, and (until a constitutional amendment in 1986) he was not permitted to succeed himself. Considering that even the brightest and most qualified officeholder cannot accomplish much in the first six months or so in office, and that a lame duck within a year or so of leaving office is in no position to wield any power, roughly half of any governor's four-year term in this state was intrinsically unproductive. Mississippi was run by its congressmen, senators, businessmen, and local authorities and police, not by its governor.

Although one of the few constitutionally mandated responsibilities of the Mississippi governor's office was emergency preparedness, Ross Barnett neglected this important task just as egregiously as had every previous Mississippi governor. When Hurricane Betsy pummelled southeastern Louisiana in 1965, and fringe effects were felt in Hancock County, Mississippi, nobody in the office of the new governor, Paul B. Johnson Jr., seemed to have the foggiest clue of what to do. Those public servants didn't, however, view their ignorance as their own fault; after all, the new governor had been in office only eight months.

For a few of the more insightful people in the coastal counties, Betsy was a call to action. If there was ever going to be any meaningful emergency planning for the coast, it would need to be initiated and implemented by the coastal residents themselves. The state officials up in Jackson were too out of touch to accomplish anything of any real value relating to disaster planning for the coast.

Julia Guice, a real estate agent in Biloxi, couldn't help but notice that the near miss by Betsy had depressed property values. Over the next four years, she and her husband, Wade, embarked on a mission—working tirelessly to draw together a network of coastal police departments, fire units, health workers, businesses, and volunteers to plan for a coordinated response to the next hurricane. Depending on who they talked to, they appealed to humanitarian perspectives, public image considerations, or more direct economic concerns. Evacuation routes and the location of emergency shelters would be publicized in anticipation of, not in response to, any future threat of a major storm. Local radio stations, television stations, and the newspapers would be linked to the key emergency management people. Evacuations would be coordinated, with extra police deputized to keep the traffic flowing. Towing companies would be on alert to quickly remove any disabled vehicles from the thoroughfares. Ham radio operators would be readied to step in to help with communications, and emergency generators would be delivered to critical command posts. After a major storm, road crews would be hustled in from unaffected inland towns to remove the debris and quickly restore the roads, while inland businesses and agencies readied their trucks to deliver food, water and medical supplies to the coast.

After the first few community leaders saw the wisdom of the idea, the momentum began to grow. It was nevertheless a monumental job to get people who had never before worked together to put aside their provincial differences for the common good. And although the three coastal counties—Jackson, Harrison, and Hancock—never did complete an effective plan to work together, at least the largest of the three (Harrison) made great strides in coordinating the emergency resources from its own cities and towns, places that included Pass Christian, Long Beach, Gulfport, and Biloxi.

By August 1969, Wade Guice was the full-time director of civil defense for Harrison County, while his wife, Julia, served as director of the Emergency Operating Center in Biloxi. Although every aspect of the plan was not yet in place, emergency preparations for the coast were infinitely better than they had been just four years earlier. Little thanks to the state government up in Jackson.

Wade and Julia followed the first sketchy news of the storm as it entered the Gulf on Friday, August 15. Although it did not appear that Camille would pose any serious threat, what better opportunity than this to test the state of the coastal emergency system they had worked so hard to create? That evening, they took turns getting on their home phone.

Saturday morning, with most of the Mississippi coast still outside the hurricane watch area, a reporter for the local *Daily Herald* phoned the Guices at home for a status report on a possible evacuation decision. After explaining that the executive committee was scheduled to meet at one o'clock that afternoon, Wade asked the reporter to request in his article that grocery stores, lumberyards, and other retail outlets voluntarily stand by to open for business on Sunday, when most of them would normally be closed. The reporter assured him he would do so. Wade smiled to Julia; the reporter had never questioned Wade's authority to make such an announcement. He handed Julia the phone, and she dialed the Red Cross, the first of a series of calls.

As the civil defense meeting began that afternoon, the hurricane warning was still confined to the Florida panhandle, the watch extended west only as far as Biloxi, and Camille's winds were around 115 miles per hour—certainly capable of inflicting serious damage to structures directly in the storm's path but hardly a serious threat to communities more than 100 miles from the warning area. On that basis, several members of the executive committee suggested a wait-and-see approach.

Wade and Julia argued otherwise. Now was the time to put all units on alert. And, unless it was perfectly clear by daybreak that Camille was headed somewhere else, a mandatory evacuation of the coast should begin promptly on Sunday morning. The function of an evacuation, Wade reminded the group, was not just to save lives. Even if the Mississippi coast got only the fringe effects of the hurricane, tree limbs would still fall, power would be out in unpredictable places, streets might be impassable, water could be scarce, and so on. The most efficient way to restore everything to normal after the event would be to have everyone without an official duty out of the way. And the only way you do that is to evacuate the coast *before* the storm strikes, when the roads are still open.

The naysayers began to reconsider. But there was still the issue of whether they might all end up looking foolish if they evacuated the coast and Camille came nowhere near. Wade moved in with his closing punch.

No matter what happened, he assured the group, they wouldn't look foolish. After all, didn't they need to know how their emergency plan worked in action? Surely there would be glitches—hopefully small ones but maybe some not so small. If they called for an evacuation that turned out in hindsight to not be warranted, they could still redeem themselves with the locals by going public with the news of how effective the drill had been. Plus, they could then use the result of the exercise to badger the folks up in Jackson to provide them more resources for emergency preparedness.

That sealed the matter. If the Weather Service should announce a hurricane warning for anywhere west of the Florida panhandle, the executive committee would immediately begin a mandatory evacuation of the Mississippi coast.

Meanwhile, the Red Cross was already preparing thirty-four emergency shelters, and thirty highway patrol units were putting their officers on alert.

Twenty-year-old Greg Durrschmidt was among the hundreds of young airmen going through specialist training at Keesler Air Force Base and nervously awaiting their orders to their new duty stations. The Mississippi coast was a place that Greg, a New York native, had quickly come to love—the long walks on the beaches, naps under swaying palms, boat trips over to Ship Island, and, of course, the wide choice of cheap watering holes around Biloxi. Too bad it was not to last forever. The best he could hope for was that he would not be shipped off to Vietnam, where each week hundreds of young Americans were being killed and hundreds more maimed for life.

Over beer at the beachfront Fiesta Club, most talk on Saturday night was about the Woodstock Festival. Nearly five hundred thousand young people, the largest such gathering in history, were congregating at a farm in upstate New York in a protest against what they regarded as a senseless war. Not that the power brokers in Washington seemed to view their motive that way. To the Nixon White House, the Woodstock happening was just a bunch of brats indulging themselves in a debauched weekend of drugs, sex, and rock and roll.

Several chums had put in for leave to attend Woodstock, had been summarily turned down, and now were curiously scarce, apparently gone AWOL. Guts or no brains? They drank a series of toasts to those lucky, or ballsy, jerks.

The topic switched to the hurricane. Several of the guys were from the heartland, and they didn't know what to expect. Greg recounted his experiences in weathering several "hurricanes" (actually their feeble remnants) on Long Island, and he assured everyone that they're nothing compared to the tornadoes in places like Oklahoma. A few tree branches fall down, he said, you lose power for awhile, and maybe the beach gets messed up a bit, but after a few days everything is back to normal. Absolutely nothing to worry about.

Someone passed a pack of cigarettes around the table. Greg had never smoked, but he took one and lit it.

"You guys notice what's playing at the drive-in?" one of the guys asked.

"No, what?"

"*Gone With the Wind.* Now do you figure that's an omen or what?" Everyone laughed.

"Hey," another airman chimed in, "you know it could be a really good omen if this hurricane blows away all the planes, so there ain't any left to take us to 'Nam!"

The joke fell flat. It was a taboo to talk about the possibility of being shipped to Vietnam.

In the wee hours of the next morning, Sunday, August 17, the outer fringes of the storm entered the range of ground radars in New Orleans and Apalachicola, Florida. Now 250 miles south of the Florida-Alabama border, Camille was still refusing to follow the meteorological rules and make the expected turn toward the north. Bob Simpson realized he needed to admit his miscalculation and revise the forecast.

In Advisory No. 13, issued at 5:00 a.m. CDT on August 17, Simpson extended the hurricane warning westward another 140 miles to Biloxi. The warning region now spanned some 280 miles of coastline from Biloxi all the way to St. Marks, Florida. The same advisory extended the hurricane watch westward to New Orleans and Grand Isle, a section of coast that includes the lower Mississippi River delta and virtually all of Plaquemines and St. Bernard Parishes. Simpson was still convinced, however, that Camille would veer north and east, and most of his staff concurred. With the storm barely seventeen hours off the Gulf Coast beaches and just twelve hours from the mouth of the Mississippi River, they collectively predicted, and announced, that landfall would be near Mobile, Alabama.

When he heard the 5:00 a.m. advisory, Wade Guice immediately set into motion the telephone chain that would mobilize everyone involved. All residents without official duties were to evacuate not just Biloxi but every place within two blocks of the coast. Teams of volunteers armed with topographical maps began tramping through low-lying neighborhoods that were considerably inland, pointing out to residents that they may be at risk even if they were not within sight of the Gulf. The corresponding civil defense coordinators of Jackson and Hancock Counties, somewhat less coordinated, also initiated their evacuation plans. Along the threatened stretch of Mississippi coast, at least eighty-five thousand people would ultimately leave their homes for higher ground and/or for points farther inland.

Wade would later reflect on how lucky they'd been. "If I had a day to pick for a storm, I would pick a Sunday, and I would pick the warning time of Sunday morning because everybody is at home. Sunday is an ideal time to have an evacuation."

He would have been justified in adding something a little less modest. Had it not been for his and Julia's proactive readiness initiative, the ensuing disaster would have claimed a monstrous toll in human life.

Robert Simpson was familiar with Mobile, still the projected landfall as of daybreak on Sunday, August 17. It was a city of some 150,000 situated at the northwest end of a bay, normally a risky configuration when a strong hurricane is approaching from the south. But here the bay's entrance is largely protected by the combination of a barrier island and a twenty-mile-long peninsula with but a five-mile gap between. The shallow waters then extend north another thirty miles before lapping at the city's waterfront. Most of Mobile, in turn, is situated on ground higher than any storm surge that Simpson figured would be likely to travel that far through the narrow inlet from the

Gulf. Even in 1969, decent roads linked Mobile with cities to the north, including Montgomery, Tuscaloosa, and Birmingham. Although Simpson was certainly concerned about the safety of the residents of Mobile, he was comfortable that he had given them a timely warning and that the officials there would know how to deal with the situation.

The coastal regions of Mississippi and southeastern Louisiana, however, were a netherland to Simpson. Yes, all of his NHC advisories and updates had been telegraphed to the National Weather Service in Washington, which in turn had teletyped and radioed them to New Orleans, the military bases, and the oil companies with offshore platforms. Simpson now realized, however, that he had no idea of how this information got from those people to the coastal residents, particularly to those rural residents living on low-lying secondary roads or on islands. He had no direct contacts with regional civil defense officials, emergency planners, the local Red Cross folks, or the Salvation Army. There was no master list of phone contacts in the files. And there wasn't the time or manpower to try to start creating such a list now, with Camille certain to strike within fifteen hours or so. A lump rose in his throat. This could easily become a replay of the Audrey disaster of 1957, maybe even worse. Hadn't anybody learned anything in the past twelve years? Hadn't he?

Simpson had accepted the NHC directorship with the confidence that he knew as much about tropical weather as anyone in the country at the time. And indeed he probably did. But although he had given deep thought to the problems of meteorological data collection and analysis, it now occurred to him that the thing he'd left on the back burner was perhaps the most important. Hurricanes in and of themselves are not disasters. It's when they catch people by surprise that such storms become disasters. How might the NHC make sure that its warnings and advisories actually reach the threatened populations? And what might be done to clarify and communicate the actual degree of the danger? Although some hurricanes may merit only minimal precautions, the way Camille was developing, she could easily become the hurricane of the century.

CHAPTER 7

EXODUS

Pass Christian, Mississippi, August 17, 1969

Ben Duckworth pulled the pillow around his ears. All he wanted to do was sleep. Who the hell was hammering on a Sunday morning?

The phone rang in the front room, and he heard Buddy pick it up. A terse muffled conversation, then the click of the receiver. Ben tossed off his sheet, yanked on a pair of swim trunks, and stumbled into the front room fumbling with his glasses.

"Gotta go to the base," Buddy said. "Hurricane warnin' and they're activatin' the med staff."

Ben rubbed his sleepy eyes, switched on the radio, and gazed out the front window. Beyond the sloped lawn of the apartment building, traffic on U.S. 90 was crawling bumper to bumper, many of the vehicles towing trailers or boats. The shallow waters of the Gulf, usually boringly placid, sported a brisk surf. But the sky wasn't particularly overcast, and, judging from the trees, the only wind was a mild sea breeze.

"Catch ya later," Buddy said as he shut the apartment door behind him.

"Yeah," Ben replied.

Ever since Ben's wife and two-year-old daughter had left for Texas last January, the apartment felt eerily empty. Buddy had moved in after the divorce, but a roommate was a poor substitute for a family, and the emptiness lingered, particularly in the mornings. Ben turned up the radio, which he kept tuned to his favorite New Orleans station. Somehow, the rhythmic horn blowing in minor keys assured him deep inside that being melancholy was okay and that life would go on.

He made himself a cup of coffee. The disc jockey briefly mentioned something about a hurricane headed toward Florida and then spun the next platter. Had Ben tuned to one of the Biloxi or Gulfport stations, he would have heard a much different story. All normal programming at those studios had been suspended, replaced by continuous announcements about the mandatory evacuation, the evacuation routes, the locations of shelters, precautionary and preparational advice, marine and small craft warnings, and updates about Camille's path. No music, no advertisements, just continuous talk about the hurricane. Those local stations had all been involved in the past four years of coastal emergency planning, and their announcers were now discharging their responsibilities with gusto.

But therein lay the first weakness of the grand plan. Many Gulf Coast residents simply didn't listen to the coastal radio stations, instead getting their entertainment and news via the New Orleans airwaves. And nobody had thought to contact those New Orleans stations to have them announce that listeners in the threatened areas should switch to their *local* stations.

The hammering and traffic continued. Ben took a last swallow of coffee, pulled a pair of jeans over his swim trunks, slipped on a pair of sneakers, and went outside to investigate. Merwyn ("Merv") Jones, the Richelieu Apartments manager, was out front, hammering plywood over the Gulf-facing ground-story windows. A bold yellow and black logo on the building reassuringly identified it as a civil defense shelter. "You busy, Ben?" Merv asked. "I could use some help."

"Sure, Merv, no problem. But I thought it's supposed to hit east of here."

"It is. This is just to keep the owners happy. They're sellin' the place to an outfit in New Orleans, but the deal hasn't closed yet, so some of the fat cats are nervous."

Ben glanced at the civil defense sign, nodded thoughtfully, and grabbed a hammer. At the very least, driving some nails might be therapeutic.

Mickey squirmed his way onto the lap of ten-year-old Richard Rose. Somehow, the black dachshund seemed to know something was amiss. Maybe the fact that nobody had fed him yet had something to do with it, or maybe it was that everyone—Richard, Don, and their parents, Fred and Juanita—was uncharacteristically hunkered around the TV first thing in the morning.

Fred motioned Juanita into the kitchen, where they could talk out of range of the boys' hearing. On the tube was a nonstop series of announcements about the evacuation. "I think we should stay," Richard said to his older brother. "It'll be fun. What could happen? Let's fill up the bathtub." He jumped up and set the bewildered Mickey on the floor.

In fact, the family had weathered Betsy at home in 1965, and that one hadn't been so bad, at least here in Gulfport. The announcers, however, seemed to be taking the current storm very seriously. As the older Roses commiserated in the kitchen, they periodically glanced through the doorway at the TV in the front room. Life had been good to them in recent years, Fred having just expanded his construction company into a new office and a full-block equipment yard on Pass Road. The family had bought a new house. They now owned two reliable cars. Should they remain to keep watch on their property or drive the hundred miles up to Laurel, where they could stay with relatives for a day or two?

Fred announced his decision: Juanita, Richard, and Don would go north, where they would be joined by their college-aged daughter, but Fred would stay to tend to the business. His collection of heavy equipment represented a considerable investment, and although he didn't expect anyone to actually steal any machinery, who could predict what might happen if a lawless environment should develop after the storm? As for Mickey, he'd be left in the garage with sufficient food, water, and chew toys and would be none the worse for wear.

If Richard had been given any choice, he would have stayed behind with his dad. He complained, cajoled, argued, and pouted. But Juanita was fully behind Fred's decision, or at least she put on that face to the boys. They were to go to their rooms and pack, load up the car, and then take Mickey's food and paraphernalia to the garage.

So it was that young Richard and Don Rose and their mother joined the long stream of vehicles heading north from the coast. Richard, still angry at having to leave, doesn't remember to this day if he said good-bye to his dad.

In New Orleans and its suburbs, no comprehensive emergency plan was in place. It wasn't even clear who in the government was in charge of making an evacuation decision (the governor never did get involved, and the mayor of New Orleans would do so only minimally and belatedly). Although most of the radio and television stations in those parts did inform their audiences of the hurricane warning, few did so in a timely or persistent manner. Some, but not all of them, advised listeners to prepare for possibly heavy winds and rain with potential flooding of low-lying areas. A few said there might be some local power outages, so it would be a good idea to stock up on batteries, candles, and canned goods. That, after all, was the standard advice. Most Louisiana broadcasters neglected to mention, however, that under no circumstances should anyone consider driving east.

In those days, nobody but a few academics had thought about the public responsibilities of broadcast media in emergency situations, let alone the problem of how those responsibilities should be discharged over airwaves that may reach ears hundreds of miles away. None of the radio or television stations around New Orleans had company policies about how to deal with hurricane advisories. Emergency broadcast content was pretty much left up to each individual program manager or whoever happened to be in charge at the time an advisory arrived. On Sunday mornings, the most experienced staffers were likely to be at home, not at the studios. And no career-minded neophyte disc jockey was about to roust a station manager out of bed on a Sunday morning with a phone call asking what to do about a hurricane advisory for another state.

The newspapers, by their very nature, were incapable of providing timely warnings. Although the main daily in New Orleans, the *Times-Picayune,* had been featuring maps of Camille's progress on its front page for the past several days, the Sunday morning edition was prepared that Saturday afternoon, when the hurricane was still expected to make landfall in Florida.

At 9:00 a.m. CDT on Sunday, with the Mississippi coastal evacuation already under way, the NHC released Advisory No. 14. Camille was holding on a steady course at 12 miles per hour, with winds of 160 miles per hour near the eye and gale-force winds extending out 200 miles in all directions. The advisory stated further, "Present indications are that the center of Camille will pass close to the mouth of the Mississippi and move inland on the Mississippi coast tonight." Hurricane warnings, for the first time, now included all of the Mississippi coast and southeastern Louisiana, including New Orleans and Grand

Isle, with gale warnings extending west to Morgan City, Louisiana. It was less than a half day before hour zero.

If the residents of rural southeastern Louisiana deeply distrusted the feds, they weren't much more trusting of the folks in New Orleans. Some might get their national news from New Orleans broadcasts, but for state and local news they consulted other sources. On Grand Isle, the local police took action well before the 9:00 a.m. hurricane warning. They got the information they felt they could trust from the oil company engineers who were shutting down their offshore rigs.

The region around Grand Isle had been washed over by a number of past hurricane-driven surges, including the 1893 Cheniere Caminada disaster. Now the region's permanent residents were mainly simple folks who earned their livelihoods from the sea. A sea they all respected when it grew angry. Roughly twenty-two hundred residents living in and around that remote place evacuated north on Louisiana Route 1, after which the police barricaded the road at Leeville, twelve miles inland.

The region west of Grand Isle stood outside the official warning area, and there would be no mandatory evacuations there. Many locals, however, refused to take chances. The three hundred people living on and around Pecan Island, including the Broussards, who had suffered such tragic losses in Hurricane Audrey twelve years earlier, immediately packed up and headed inland. As far west as Cameron, two hundred miles beyond the western limit of the official warning area, fifteen hundred people heeded the terrible lessons they'd learned from Audrey and evacuated voluntarily.

Although the U.S. Army Corps of Engineers had no jurisdiction over the civilian preparations, the administrators of the New Orleans District of the Corps of Engineers well knew that if there should be a disaster, the responsibility would fall on them to clean up a big part of the mess. The corps had learned some grave lessons about flooding and infrastructure failures after Betsy in 1965. At noon on Sunday, the district activated its Emergency Operations Center. Generators were tested, and a backup communication system was established, linking headquarters with all personnel in the field. All tidal and river gauge information was to be continuously relayed to headquarters, which would be in contact with the New Orleans civil defense authorities (regardless of the fact that the latter seemed somewhat lackadaisical about the corps's concerns). And word was sent to Washington to place helicopters and amphibious vehicles at the disposal of the district on the heels of the storm.

There was one more matter for the corps to address. A new pumping station was under construction on the south shore of Lake Pontchartrain, and a temporary levee had been built lakeside of that site to protect the project against high tides. If the lagoon should top that levee, the construction would be destroyed in the surging water, and the hammering effect against the permanent levee would seriously stress it as well, threatening hundreds of homes. The engineers thought it over and came to the solemn and expensive conclusion that the levee would need to go. That afternoon, after removing the mechanical equipment, they breached the temporary levee and let the waters of Lake Pontchartrain flood the site. The sole purpose of this drastic action was to reduce the hazard to several blocks of city residents.

Could it be that the Louisianans were wrong and that the feds really *did* care about them? During the hardships that would follow, few locals would think about it in such terms. The boys at the New Orleans District of the Corps of Engineers, after all, lived in *Louisiana,* not in Washington. And when you live in Louisiana, that's where your heart's going to be.

That Sunday afternoon, about eight hours before Camille would strike Mississippi, another air force C-130 penetrated the eye and measured a barometric pressure of 26.62 inches of mercury. Camille, in other words, was continuing to intensify. Copilot Robert Lee Clark would write in his official report:

> Just as we were entering the wall cloud, we suddenly broke into
> a clear area and could see the sea surface below. What a sight!
> No one had seen the wind whip the sea like that before.
> Instead of the green and white splotches normally found in a
> storm, the sea surface was in deep furrows running along the
> wind direction. The velocity was far beyond the descriptions
> used in our training.

Getting a numerical value for the wind speed, however, was not an easy matter. The cockpit airspeed indicator tells how fast the craft is moving through the air, not how fast that air itself is moving over the ground. The main technique for finding a hurricane's wind speed was to "fly a box," which involved flying, say, east for two minutes, then north for two minutes, then west, and then south for equal times. In still air, this brings the plane back to its starting point. If the air is moving, the box fails to close on itself; the faster the wind, the greater the discrepancy. The mathematics, although not exactly simple, is reasonably straightforward.

Using this method, and with technical assistance radioed from the NHC, the C-130's navigator computed Camille's winds at an astounding 190 miles per hour. The crew had to terminate further observations when the plane lost an engine. Their exit to Ellington Air Force Base near Houston was harrowing but successful. No further flights would be made into Camille.

Robert Simpson read the figures in astonishment. Never before had anyone measured a steady 190-mile-per-hour wind speed in a hurricane. Hurricanes of 120 miles per hour or less had been known to be deadly, and a 190-mile-per-hour wind packs two and a half times the wallop of a 120-mile-per-hour storm. Simpson's staff checked and rechecked the data and the computations, but the results were consistent—190 miles per hour, at least at the altitude of that recon plane.

Simpson included the new wind speed figure in Special Advisory No. 16 (issued at 3:00 p.m.), along with the news that Camille was now traveling at fifteen miles per hour—which meant that she would strike the coast slightly earlier than previously expected. Already, the winds were approaching hurricane force at the mouth of the Mississippi River. The projected landfall was shifted westward once again— to Gulfport, Mississippi.

Something etched deep in Bob Simpson's subconscious exactly fifty years earlier began tugging at his tired brain. He returned to Chester Jelesnianski's storm surge computations. Should he raise the storm tide estimate? It already stood at fifteen feet; would anyone believe a prediction of twenty-five? Simpson pored over the sea charts for the region; except for the shipping channels, most of the soundings were seven feet or less. Was it credible that a twenty-foot bulge might develop on a body of water only seven feet deep? Where would all that extra water come from? And how would it get there that fast?

After all, Camille herself was now moving forward at fifteen miles per hour, and ocean currents don't travel at anywhere near that velocity.

On the other hand, what if Jelesnianski's predictions were right? A twenty-five-foot storm tide would generate horrendous currents, and even sizeable ships could be carried inland. Any structure within several blocks of the shore would be battered beyond recognition. As for anyone unwise enough to remain near the coast, survival would be virtually out of the question. After rounding the surge figures downward by a few feet, Simpson added them at the end of the 3:00 p.m. bulletin:

> THE FOLLOWING TIDES ARE EXPECTED TONIGHT AS
> CAMILLE MOVES INLAND . . . MISSISSIPPI COAST
> GULFPORT TO PASCAGOULA 15 TO 20 FEET . . .
> PASCAGOULA TO MOBILE 10 TO 15 FEET . . . EAST OF
> MOBILE TO PENSACOLA 6 TO 10 FEET. ELSEWHERE IN
> THE AREA OF HURRICANE DISPLAY EAST OF THE MIS-
> SISSIPPI RIVER 5 TO 8 FEET.

This advisory, however, did not mention Pass Christian or Bay St. Louis or any specific places in Louisiana. From this unfortunate omission, many readers would conclude that any flooding west of Gulfport would not exceed eight feet.

Louisiana State Route 300 terminates at Delacroix Island, an obscure cheniere in St. Bernard Parish. Most of that place's residents have surnames like Gonzales, Martinez, and Molero and are descended from the 2,010 immigrants who arrived in this region from the Canary Islands in 1778–79. Even today, these locals still refer to themselves as "Islenos."

Delacroix was the site of the so-called Trappers War of 1923, triggered when young Judge Perez and his uncle concocted an ill-conceived scheme to require fur harvesters to pay license fees. On the morning of November 16, 1923, some four hundred armed and angry locals were gathered there to discuss this extortion attempt when an oyster lugger, the *Dolores,* chugged into view. Sent by Judge Perez to pacify the rebellious trappers, the boat brimmed with armed deputies, two mounted machine guns flanking its helm. Within minutes, the shooting started. Coroners later counted seventeen bullet holes in the body of one of the machine gunners; the other, attempting to surrender, was shot three times but lived. Virtually every man on board was shot at least once. The *Dolores,* riddled with bullet holes and her fuel tank punctured, ran aground. The trappers disarmed the surviving deputies, jammed them into a couple of trucks, drove them up to New Orleans, dumped them off at a hospital, and told them to warn Judge Perez not to ever harass them again. With that, the violence ended and life in the quiet St. Bernard marshes went on as before.

When Betsy ripped through Delacroix in 1965, the surge submerged the whole cheniere, destroyed most of the homes, beached and otherwise damaged every boat, and drowned several dozen residents. Now, barely four years later, most of the dwellings were trailers or modest cabins built from salvaged lumber, all strung in a single line on one side of the main road, their backyards protected by a small levee no higher than four feet. The other side of the street was paralleled by a canal dotted with tattered-looking marine repair facilities and piers berthing rusty but functional shrimp boats.

At the news that Camille was headed their way, most of the Isleno men sent their families to the fire hall in Reggio, six miles farther inland. That village hadn't flooded during Betsy, so surely that was good enough. Reggio, however, was also under a local evacuation advisory, and as refugees poured into the municipal buildings from the outlying areas, most of that town's own residents were fleeing to the north. Although the official post-Camille reports would document that sixty-seven hundred people evacuated the low-lying regions of St. Bernard Parish that afternoon, in fact many of them simply relocated from one low-lying region to another.

Meanwhile, most of the men of Delacroix Island remained behind on their boats, ready to adjust the mooring lines and to take other measures to limit the damage when the expected storm surge swept in. Within a few hours, those men were to get a strange surprise.

One hundred and sixty miles to the east, local officials were still operating under the assumption that the storm might strike the Florida panhandle. The four thousand inhabitants of Santa Rosa Island were evacuated as of mid-afternoon, and twenty-four hurricane shelters on the mainland around Pensacola bustled with activity.

Meanwhile, the Alabama coast was making its own preparations. At the mouth of Mobile Bay, the *Ocean Explorer* was about to start jacking up an Odeco drilling rig—a time-consuming and delicate operation—when the captain received instructions from headquarters to delay the operation. Yet to allow the rig to float there during the storm would be risky; it would be uncontrollable and could easily be swept aground or, worse, crash into one of the many stationary platforms and destroy a great deal of expensive equipment. After commiserating about the situation, the Odeco engineers sent the *Ocean Explorer* out into the Gulf and then westward toward the eastern coast of Louisiana. That would put the rig several hundred miles west of the hurricane. Or so they figured.

Camille, however, didn't oblige. The platform was being towed, it turned out, in exactly the wrong direction.

Around 3:00 p.m., a tractor trailer capsized and blocked Route 49 near Reed Road in Gulfport. Although tow trucks had been placed on alert to remove disabled vehicles from the thoroughfares, an overturned eighteen-wheeler on one of the three main routes inland was a more difficult problem than a mere stalled car. To remove the tractor trailer would have required stopping traffic completely for an hour, maybe longer. The authorities decided to simply let the truck lie where it was and to direct traffic around the obstacle as best they could. The evacuation, almost bumper to bumper already, jammed up further.

The traffic snarl stretched down the beach road past the Richelieu. As the plywood nailing was nearly complete, or at least as much of it as manager Merv Jones figured needed to be done, Ben noticed an elderly resident waving at him. Jack Matthews, whose wife was in a wheelchair, motioned for Ben to come over. "You and your friends stayin'? If you are, I'd sure appreciate it if you could look after Mrs. Matthews and me." Ben assured Mr. Matthews that he'd be glad to if he stayed, but he had yet to decide.

Nearby, also surveying the confusion on Highway 90 were newlyweds Rick and Luane Keller. Rick, an engineer at the Mississippi Test Facility, was a Massachusetts native who had seen some fierce nor'easters. Could a hurricane be any worse than those violent New England storms? Rick asked Ben for his opinion. Stay or leave?

Merv Jones sauntered over, his T-shirt drenched in perspiration, hammer still in hand, a reassuring smile on his face. What does a prudent person do when he faces the unknown? He shares what he knows, or what he thinks he knows, with those around him, and he reciprocally listens to what they have to say in return. The Richelieu, Merv explained, had a steel skeleton. And although that would turn out to be an unfortunate piece of misinformation, to engineer Rick Keller and later to Ed Bielan, another NASA engineer who would also remain, it meant that the structural integrity of the building was not an issue. There might be superficial damage, broken windows, and even flooding on the ground floor, but there are few destructive agents on the planet that will bring down a steel-framed building.

As for the prospect of flooding, nearby Henderson Point had been under nearly fifteen feet of water during the disastrous hurricane of 1947, but even in that storm the flooding at the higher site of the Richelieu was only three feet. That would be enough to damage the current ground-floor units but nowhere near high enough to reach the second floor, let alone the third. And that 1947 storm surge had set the standing record for hurricane flooding along the entire Gulf Coast.

Merv and his wife Helen were planning to ride it out. "There's some empty apartments on the third floor," Merv offered. "Y'all are welcome to stay up there if you want to. Meanwhile," he grinned, "y'all can help me move some furniture upstairs."

Rick looked lovingly at his young bride, her hazel eyes sparkling, her strawberry blonde hair tossing in the breeze. Luane was a pretty girl—some would say beautiful—and she was Rick's pride and joy. "We should get some flashlights and things," she said with a smile.

Ben still wasn't sure. But with traffic as heavy as it was and these folks still needing help, this wasn't the time to leave. He announced that, as a precaution, he was going to move his brand-new Chevy Impala to higher ground and he'd be back shortly. With that, he trotted to the parking lot and drove off through the residential backstreets of Pass Christian, eventually finding a protected parking spot next to the brick rear wall of the Catholic church.

As he started to walk back, a police officer called to him. A number of other folks were looking for high ground to leave their vehicles. Would Ben stick around and shuttle those people home after they parked? The cop would reserve Ben's own parking space for him while he did so and then would drive him back to his apartment. Thus deputized as a valet, Ben shuttled eight car owners home and then nested his own shiny gold Impala next to the ostensibly protective church building.

Around six o'clock that evening, the officer dropped Ben off at the Richelieu, never suggesting that it might be imprudent for him to remain there. There was now a steady drizzle, but the furniture moving was still in progress. The small group of sweaty men on the stairs, including Merv and Rick, cheered when they saw Ben. They had begun to think Ben had abandoned them.

Sometimes there are reasons for doing things that can't quite be articulated but that are no less valid for remaining unexplained. Ben wiped the rain off his glasses, set them securely on his nose, and began carting furniture up the Richelieu's concrete stairwells.

The question of why some folks stay while others leave has confounded emergency planners for decades. Why did thousands of people remain on the Mississippi coast when they knew Camille was on the way? And why has such behavior repeated itself during one hurricane after another?

Despite the best efforts of authorities to insist that an evacuation is "mandatory," law officers are empowered only to empty out public places and commercial establishments; police cannot force out people who hunker down in their own homes. Some people refuse to evacuate because shelters and hotels won't accept pets, and they will not abandon these beloved companions—a part of their family—to an uncertain fate. This is particularly relevant to the elderly whose pet may be their *only* companion. People died in Camille because they wouldn't leave their pets behind.

Every individual has his or her personal threshold for risk acceptance, and everyone has a different set of experiential and informational bases to draw from in reaching a personal decision. As a result, a hurricane evacuation is not exactly a group behavior; it's a superposition of thousands of individual decisions, each of which reflects what that particular man or woman knows, when they know it, and how they feel about it at the time.

In Camille's aftermath, the Weather Service contracted with Kenneth Wilkinson and Peggy Rose of the Social Science Research Center at Mississippi State University to conduct a study whose report would be titled *Citizens' Responses to Warnings of Hurricane Camille.* Those researchers and their field-workers interviewed 384 residents of Harrison County to ascertain any special characteristics of those who left versus those who stayed. Of the leavers, more than 70 percent did so within seven hours prior to the hurricane's arrival, 89 percent used a personal vehicle (versus 8.3 percent who walked and just 1.1 percent who used a public vehicle), and more than 80 percent evacuated with family members. About one-third went to public shelters, and a roughly equal number went to the homes of friends or family on the coast, which in some cases actually placed them at greater risk. Only one-fifth of the leavers moved inland any significant distance.

Almost one in four people remained. The stayers were predominantly male, slightly better educated than the leavers, and somewhat older (a median age of 54.2 years versus 48.1 years for those who left), and on average they owned more expensive homes on higher ground. They were slightly more likely than the leavers to have had a previous personal experience with a hurricane. They were more likely to listen to out-of-town radio stations, and (curiously, considering their education) they were *less* likely to be able to state the difference between a hurricane watch and a hurricane warning.

Many, if not most, of the stayers expected that the worst part of Camille would strike elsewhere. Some had no concept of the threat from a storm surge; others figured that their homes were high enough and otherwise safe enough. Some were too tired or too sick to leave. Some were worried about protecting their belongings. Others were concerned about the cost and inconvenience of trying to find shelter inland. There were those who didn't own cars or whose cars weren't reliable enough for a long trip. A few, like Ben Duckworth, were too busy until it was too late to leave safely. Some simply procrastinated too long for no important reason. There were a few, of course, who thought that being in a hurricane might be a memorable experience, something to tell their friends about and someday their children and grandchildren. And then there were those who had no choice to leave: prisoners and members of the military.

An unlikely prisoner, a young college student named Lucretia, had a summer job not uncommon in the 1960s: trekking through neighborhoods to conduct a survey of the residents' reading habits. Her employer, an encyclopedia company, would then follow up with sales pitches to prospective customers. Lucretia had three things going against her: (1) she was half black and half American Indian, (2) her British companion was white, and (3) they were conducting their survey in the small town of Brandon in central Mississippi. Almost predictably, in the racial atmosphere of those times, they were arrested on trumped-up charges, thrown into the Rankin County jail, and given no access to outside communication—not even a single phone call. Sally, Lucretia's white companion, was taken to a cell upstairs while Lucretia was put in a dingy, filthy basement lockup, separated from a cell full of men only by the steel bars, with no privacy for even the toilet. It was in that jail that the two women would separately weather the triple ordeal of incarceration, humiliation, and Camille. For a week they would be without edible food and minimal water, the butt of threats and obscenities and worse. "You name it," according to Lucretia, "and it happened."

In Gulfport, Mayor Philip Shaw acted more responsibly. Around 9:00 p.m., with the windows of the police station already rattling, he made the decision to free everyone incarcerated in the jail. Every single one refused to leave. Even the next day, when those prisoners were given official clemency, two of them would still refuse to leave.

Greg Durrschmidt, one of the young airmen at Keesler Air Force Base, was entertaining a visiting friend that Sunday, and as they took a nickel tour of the area they tuned the car radio to a New Orleans station. The only news that particular DJ had to report was about the Woodstock Festival in upstate New York, now in its last day. As they hopped from one bar to another, the increasingly gloomy sky gave way to a steady drizzle. Greg still vividly remembers the traffic snarl on Highway 90 and the plywood going up over windows all along the beach. The Ship Island ferry was missing from its dock, as were most of the other boats he usually saw moored there. With the first outer rain squalls whipping through and the bars closing, they returned to the base. There, the flight line was weirdly deserted, and even the dozens of training aircraft were gone.

The official word from the base commander reflected stale information. Because Camille was going to strike a considerable distance to the east, the airmen would not be evacuated. Instead, they were restricted to their quarters until further notice, which would be at least until morning. Under no circumstances were they to leave their quarters.

And then there were those who, despite their best efforts, made every effort to leave and were just plain unlucky.

The *Rum Runner* was a beautiful forty-five-foot Biloxi schooner, built of cypress in 1929, that for several decades had been used to smuggle liquor from Havana to the Florida Keys. For Earl Hover, age forty, owning and living on such a boat had been his lifelong dream, and now it had been fulfilled. On August 10, however, during a sailing trip with three friends from New Orleans, the old diesel engine got cranky. They anchored at the Broadwater Beach Hotel Marina in Biloxi, unsuccessfully tried to make the mechanical repairs, and then were driven home in embarrassment by one of their wives.

The following Sunday morning, the four men—Earl, Frank Murray, Ronald Durr, and Durr's father-in-law, Charles Dussel—returned to Biloxi to retrieve the boat and complete their trip, even as most of the other traffic was headed out of town. It wasn't that they didn't know about the hurricane; it was that they did. Camille was forecast to strike to the east, but with the possibility of gale-force winds extending as far as the Mississippi coast, leaving the *Rum Runner* at the Broadwater Marina did not make good sense. Based on all reports, by heading west they would be sailing *away* from Camille's storm track. And, according to all calculations, they had plenty of time to reach a berth in Lake Pontchartrain before dark. So it was that at 10:30 a.m. on Sunday, August 17, with a Coast Guard boat towing them out of the harbor to where the sails could be raised, the *Rum Runner* began its last voyage, hauling a fourteen-foot dinghy.

There are two ways to enter Lake Pontchartrain from the Mississippi Sound, and both involve crossing part of the shallow (seven to ten feet) lagoon called Lake Borgne. The most direct way in is through the narrows of The Rigolets with its two drawbridges, a tricky business for a boat under sail power alone. The alternative is to sail farther southwest, round Alligator Point, and enter the somewhat shorter and more navigable Chef Menteur Pass. That would be the plan. Under sail power alone, however, they would get only one chance to tack in, so they had to get it right the first time.

Unbeknownst to those four Sunday sailors, an invisible east-moving tidal current was hindering their progress. The sun was already setting when they passed The Rigolets. From there they tacked south, out of sight of land. None of the four had taken this route before, and they argued about whether they had gone out far enough to tack back toward shore. Durr said no; the others outvoted him. They tacked in, and in the waning light they realized that Durr had been right—Alligator Point lay a hundred yards dead ahead, blocking their access to the pass on their present tack. Even now, they had no inkling about the true reason for the shortfall: Camille was affecting tidal currents hundreds of miles from her center and way in advance of her winds.

They dropped anchor, let out a long length of anchor line, and took down the sails. They had no choice but to spend the night where they were. At least the *Rum Runner* had four bunks and a head, so that wouldn't be so bad. Everyone would be rested, calm, and friendlier by daybreak. And the upside was that they were well out of the path of Camille.

Or so they thought.

CHAPTER 8

TROUBLED WATERS

At daybreak on Sunday, August 17, 1969, some thirty freighters and tankers were already queued up outside the offshore sandbars, waiting their turns to enter the Mississippi River's two deep-water passes. Dozens of additional ships would arrive throughout the day as their masters sped full ahead to gain haven before the sea grew treacherous. Although the entrances to the passes were always bottlenecks to shipping, today those waters would be especially crowded and confused.

It is near the edge of the continental shelf, just outside the shifting bars and mudlumps, that a pilot is taxied out to a waiting vessel, boards the ship via a rope ladder slung over its side (a so-called Jacob's ladder), and guides the ship north to the Pilot Town anchorage, where he disembarks and is replaced by another pilot. It should be noted that no pilot ever puts his hands on a ship's wheel, nor does he issue any direct orders to anyone. His job is to stand on the bridge beside the officer on duty and advise him where he should or should not point the vessel. According to many pilots, the hardest part of the job is climbing that swaying ladder, which is typically around thirty feet in length. Pilots have been known to lose their grip, particularly in brisk winds and choppy seas, and more than a few have not lived to tell that tale.

Louisiana pilots are not government employees but rather belong to one of four guilds that hold state-endorsed monopolies on piloting—organizations whose financial operations, like so many things that merge politics with money in Louisiana, remain cloudy today and were even murkier in 1969. In 1969, a river pilot earned about $50,000 per year, or roughly five times the national median family

101

income; in 2004, their reported earnings were $342,000 per year, while the Calcescieu canal pilots in the guild at the western end of the state earned an average of about $500,000. Given the pay, it is hardly surprising that pilots cling to their jobs as long as possible. In fact, several of the pilots who drowned in the river or the Gulf were in their seventies.

Although the official dogma in 1969 was that a pilot entered the guild only after a long and grueling period of apprenticeship, more than one well-connected tyro bragged over a beer that he'd shown such a natural talent that the whole training process had taken him only a few months. Meanwhile, an obstacle course of hazy requirements—some written, some unwritten—has always confronted any outsider aspiring to join the pilot guilds. Back in the 1960s, the chance of an interloper getting into this line of work was about as good as that of an armadillo sprouting wings, and until 1990 no apprenticeship ever went to anyone other than a white male with the right family ties. It was not until the spring of 2004 that the Louisiana State Legislature enacted a statute providing for substantive regulation and oversight of the activities of the state's four pilots' associations.

To visit the Head of Passes, you need to charter a boat near the end of the road at Tidewater, a dozen miles upstream. After navigating three miles of canals and waterways lined with tank yards, workboats, and drilling equipment, you enter the river through an inlet known as "the jump." The Mississippi River is about three-quarters of a mile wide here, turbulent and choppy, with swirling eddies and erratic currents blanketed with flotsam. Occasionally, a waterlogged tree or chunk of building timber surfaces, swirls around for a few seconds, and then disappears into the depths. Freighters and tankers churn past, both northbound and southbound, their superimposing wakes adding to the confusion and creating a bumpy ride for any small vessel. There are virtually no pleasure craft on this section of river; pilots can't easily see them from the bridges of the big ships, and there are many safer routes for small boats to use to get to the Gulf.

The Main Pass, which hasn't been navigable for centuries, veers off to the northeast. Next comes a sprawling marine fuel depot. A short distance farther south, the Pilot Town docks appear. Several white workboats are typically moored here, the word *PILOT* emblazoned in bold red lettering on their superstructures.

Pilot Town's main street is an elevated boardwalk about six feet above the spongy ground, and all of the occupied buildings are perched yet higher on pilings. The most prominent structure is the sturdy old bunkhouse, replete with a large sign demanding, "Quiet. Pilots Asleep." Everything within sight of the landing areas is neat and well maintained. Only a considerable distance from the docks does one notice the decaying and overgrown ruins of other former buildings, artifacts of years gone by when Pilot Town was a family community.

The Lower River, August 17, 1969

The mushrooming number of incoming ships overwhelmed the twenty or so bar pilots on duty that morning. Already the Gulf was getting choppy, and, according to the marine forecasts, conditions were expected to deteriorate considerably by late afternoon. From the height of a ship's bridge, a driving rain can reduce visibility to where you can't even see the river below. Couple that with hurricane-force winds that can impede the rotation of radar antennas while crabbing a ship in unpredictable directions, and it can become impossible for a pilot to judge where the vessel is heading.

The Branch Pilot Association's chief engineer had a decision to make. Around noon, he issued an advisory to all shipping and maritime interests that bar piloting activities would be suspended until after Camille had passed. The waiting ships—the exact number is unknown, but it seems to have been around fifty at that time—dropped their anchors as their crews cleared and secured the decks in anticipation of the storm. The captains of those cargo vessels and tankers would be on their own when Camille struck, attempting to hold steady in unfamiliar waters with unpredictable currents in high winds under conditions of near-zero visibility. In hindsight, the outcome may have been no worse if all those ships' captains had simply broken the law, taken their own chances, followed one another up the river without waiting for pilots, and paid the fifteen-thousand-dollar fine for having done so. If some of them were to run aground in that process, well, many of them were about to be run aground anyway.

Although the New Orleans customs house kept meticulous records of ship arrivals and departures at the city's wharfs, there was no equivalent of an air traffic control system that monitored the whereabouts of merchant vessels in the river or the Gulf. Amazingly, there was not even a central depository of records on shipwrecks and groundings. The U.S. Coast Guard kept files only on those specific incidents that involved Coast Guard operations. According to state law, the Louisiana Department of Transportation (LaDOT) was supposed to keep records of shipping accidents on navigable waterways within the state; unfortunately, however, none of the bureaucrats at LaDOT seem to be aware of this even today. What happened in Louisiana waters during Camille was later reconstructed by the U.S. Army Corps of Engineers, whose official report would state, "Marine losses were derived through contact with insurance underwriters, marina operators, local boat owners, and estimates made in the field." Hardly the kind of definitive sources one might expect.

The pilot houses at the ends of the South and Southwest Pass jetties (respectively named Port Eads and Burrwood) were battened up and evacuated by early afternoon on the day of the storm. Some of the pilots headed home to the New Orleans area, while a few went farther north. Ten of the pilots, along with the Branch Pilot Association's chief engineer, hunkered down in the bunkhouse at Pilot Town, twenty miles upriver from the Gulf. None of those men had any way of knowing how many ships would be forced to weather the storm out in the Gulf; the captains had quit radioing in after the pilot's association announced it was suspending operations.

Ships seldom anchored at Pilot Town; pilot transfers were usually made while the ships were in motion (albeit slowly). Even refueling took place while a ship was under enough power to buck the current. That Sunday, however, things were different. A dozen freighters would need to drop anchor there to ride out the storm in the river, including the 9,978-ton *Cristobal,* the 10,167-ton Italian vessel *Sirio,* and the 16,657-ton Liberian ship *Jela Topic.* Meanwhile, perhaps a hundred other vessels of various sizes and descriptions, from tugs to dredges to work barges, were making preparations to anchor or moor along the lower fifty miles of the great river.

The *Buffalo* was a thirty-foot wooden boat well suited to light commerce on the river: eight-foot beam, shallow draft, inboard engine, screw propeller, and a canopy over the open helm. Nobody knew when or where it was built. Joseph Jurjevich, a Croatian immigrant, had bought it at a New Orleans wharf in 1918.

By the time the wavelet of Croatians arrived in the early 1900s, the most desirable lands of the Louisiana Purchase had long been settled by earlier immigrants. Just south of New Orleans, however, all the roads terminated and the wetlands began. Although numerous Isleno families in St. Bernard Parish had been eking out modest livelihoods in the eastern parts of those swamps, a broad expanse of soggy real estate remained untouched by civilization. Here, it was still possible for an ambitious immigrant to derive a living from the natural bounties of the marshes and estuaries.

To gain access to that region, however, one needed a boat. The *Buffalo*'s previous owner had bought it only for its engine, which he removed before putting the immobilized hulk up for sale. The newly arrived Jurjevich recognized opportunity when he saw it, and he scoured the New Orleans blacksmith shops and marine yards for a gasoline motor (a used one, of course), had it installed in the *Buffalo*, negotiated a workable payment arrangement, and chugged off happily down the river to the little village of Olga on the east bank—where all of the homes were perched on stilts and most of the men harvested oysters for a living. There, the young Jurjevich joined a community of a few hundred other immigrants from Croatian Austria—folks with names that are still common in this region: Cvitanovich, Dekovic, Jurisic, Pavlovich, Petrovich, Popich, Rusich, Squarsich, Taliancich, Tesvich, Vujnovich, Yankovich, Zuvich, and so on.

With his own hands, Joseph Jurjevich replaced most of the *Buffalo*'s bent-oak planking with more durable indigenous sawed cypress. Soon he was supporting himself by shuttling cargo, oysters, and passengers between lower Plaquemines Parish and New Orleans and by delivering mail three times a week along the lower river. He managed his resources well, and before long he had enough money to build himself a modest house in Olga and to marry a young Croatian woman.

From his earliest days in politics, Judge Leander Perez had taken a liking to these hard-working Croatian families, and they to him. He may even have been the source of the original loan for the *Buffalo*.

Perhaps Perez's admiration had something to do with the fact that the Croatians were masters of the labyrinth of swamps in the region, where smugglers and bootleggers inevitably crossed paths with oystermen and trappers, in a lawless environment where expectations of reciprocal favors might easily develop. Although there is no hard proof of just who Perez's confederates were in his early smuggling and bootlegging operations, it is not unreasonable to speculate that some of the Croatians may have played a role. Regardless of the reason, when the enterprising Joseph Jurjevich became a father in 1930, he asked Perez to be his son's godfather. The little guy's name, of course, would be Leander.

Over the decades, the *Buffalo* became a common sight on the lower river, functioning variously as a packet, a hearse, a water taxi, and a pleasure craft. On occasion, Judge Perez chartered the vessel for his own purposes, including hunting trips and other excursions whose purposes weren't always apparent. In 1969, the old boat was still river worthy and still in the Jurjevich family.

Many things, of course, had changed by then. Olga had become a virtual ghost town, most of its buildings having succumbed to floods and tropical storms over the decades. Only the elder Jurjeviches and a handful of other old-timers still lived there. Leander Perez's godson, Leander Jurjevich, had moved across the river to Boothville, where he got himself a job working for the parish under the supervision of Luke Petrovich, built a home, and began to raise his own family. He kept the *Buffalo* moored two miles north of his home at the long-decommissioned Fort Jackson, using the boat to ferry supplies over to the old folks in Olga, which was still inaccessible by any land route.

When Hurricane Betsy flooded the parish in 1965, Leander Jurjevich and his family lost all of their belongings except for the boat. As with the rest of Betsy's many victims, their insurance company paid only about 10 percent of the home's replacement cost. How Leander felt about that can only be imagined; today he just shrugs and declines to say much about those ancient feelings. And at the mention of Judge Perez, he merely smiles, a mischievous twinkle in his eye.

By 1969, the Jurjeviches' new two-story home was complete. The ground floor, used mainly for storage, was built of steel-reinforced concrete block. The sills of the wood-frame second floor were tied to those block walls with thick lag bolts set in poured concrete. The extra cost of doing all that was offset by the Jurjeviches' peace of mind that they would never again be stiffed by an insurance company. They were seemingly well prepared for the next hurricane.

Evacuating in the face of Camille had not really been considered. Leander's brother, Joseph Jr., had come home from the air force to visit the old folks and to do some fishing. The eldest Jurjevich, however, had heard the weather forecast and insisted that nobody was going fishing that day; instead, he and Mama would leave Olga and spend the night at Leander's place. The brothers ferried their parents across the river on the *Buffalo* and tied up at Fort Jackson, and everyone congregated at the solidly reinforced Jurjevich home in Boothville. Leander never figured that the storm would amount to much; the worst of it was supposed to strike far to the east.

Late that afternoon, with the wind beginning to gust and an intermittent drizzle already in progress, they heard an announcement that Camille was going to strike farther west than previously predicted. Leander and Joe Jr. decided to drive back to the levee to make sure the *Buffalo* was secure. Leander's parents; his wife, Margarethe; and his children, Lea Jr. and Charmain, remained at home.

The brothers drove past the ramparts of the old fort, drove up onto the dike, and parked next to the *Buffalo*. The southern sky was the color of swamp mud, visibility downriver was a few hundred feet and deteriorating, and a fierce chop had developed. This was obviously going to be a nasty storm. In a tug-of-war against strangely swirling currents, they resecured the *Buffalo*'s hawsers.

It was too late to make it back to the house. Lacking an enclosed cabin on the *Buffalo*, they dashed to the *Sally Kay*, a boat used for ferrying workers to and from the offshore oil rigs. They climbed aboard just as the crew began playing out the mooring lines. Then came a jolt and an abrupt surge in the river. As Leander and Joe Jr. watched in horror, the lines popped on the *Buffalo*, and it disappeared over the levee. Moments later, Joe's car followed.

Considering that he'd built his house to withstand a hurricane and an eight-foot flood, Leander felt sure his family would be okay. He just hoped they wouldn't be too worried about his and Joe's failure to return. They'd get themselves home as soon as the storm passed.

The thing that was going to be most painful, he kept thinking, was breaking the bad news to his dad about the demise of the *Buffalo*.

The first spot to experience Camille's fury was Garden Island, a mucky low-lying swath of land just east of South Pass. The Freeport Sulphur Company had built a cluster of bunkhouses there, along with industrial buildings and a barge terminal for its mining operations. As the foremen evacuated the light equipment and most of the workers, they solicited volunteers to remain through the hurricane. That storm crew's job would be to radio a preliminary damage report and a list of immediately needed repair materials back to company headquarters in the town of Port Sulphur.

Lured by the prospect of a bonus in their next paychecks, a handful of eager young men agreed, expecting to be safe in the reinforced concrete power-plant building. As the last boat departed, they brewed some coffee, took out a couple of decks of playing cards, and settled in. The rain was just beginning.

Around 6:30 p.m., an unprecedented sixteen-foot storm surge swamped the building's lower two floors. The generators, of course, immediately failed. The men scampered to the third floor and huddled in an interior room with water swirling around their ankles, praying through the following terrifying hours that the sea would rise no farther.

They were lucky; they were in the milder left front quadrant of the hurricane. Even so, all of the production facilities and living accommodations at the Garden Island site were obliterated. The only structure that withstood the battering was the generator building, to whose integrity those men owed their lives.

About twenty-five miles to the north, five meteorologists had decided to spend the night at the weather observatory in Boothville. Storms are in such men's blood, and the opportunity to study this hurricane from the inside was not an occasion to miss. They didn't think of their action as foolhardy, for the observatory stood on pre-stressed concrete pilings with its main floor twelve feet above the ground.

Soon came a disappointment: after recording a sustained wind of 107 miles per hour, the anemometer failed, as such instruments seem to have a habit of doing just when they are being called upon for important duty. Then came a rude surprise: a sixteen-foot flood, even though they were two dozen miles from the Gulf. As the windows were shattered by flying debris, the men took shelter in an interior electronic shop, waist-deep in water. The time then was about 8:00 p.m.

The egress of the Mississippi River into the Gulf of Mexico, all two billion cubic feet of it *per hour,* depends on the simple physical law that rivers flow downhill. After about 7:00 p.m. that evening, with the Gulf rising sixteen feet higher than the river, the slope was reversed. In its confusion, the lower end of the massive stream swirled into great eddies and whirlpools, and when it settled down it was running backward and the sea was pouring into its mouth. Camille, in only a glancing blow, had managed to reverse the flow of the mighty Mississippi.

How much of the river ran backward is well documented from the tidal and salinity gauge data collected by the Army Corps of Engineers. The result is astounding. During the hurricane, the river reversed its flow from its mouth to Carollton, a distance of about 120 river miles. At New Orleans, where the river surface was four feet above sea level in the morning, by late that evening, it was eleven feet above sea level—a rise of seven feet, virtually none of which was due to the rain (regional rainwater flows *away* from the river, not into it). Perhaps equally, if not more, astounding, the effects of the backup were documented as far inland as 50 miles north of Baton Rouge, or some 295 miles upriver from the Gulf of Mexico. In this upper section above Carollton, the river did not actually reverse its flow, but its current was slowed so much by the downstream backup that its level was nevertheless forced to rise.

At Pilot Town, with the bunkhouse shuddering in the wind, a row of five-gallon jugs of water walked their way across the floor no matter how they were arranged. At least those unruly water bottles gave several of the men something to occupy their minds and their hands. Others peered through the narrow slats of the shuttered windows at the lights on the freighters at anchorage. There, something weird was happening. One thing every pilot develops is a keen sense of motion and its direction. Sighting the ships' lights and the piers in the foreground, the men agreed that it *looked* as if their bunkhouse was moving, yet they all agreed that in fact it wasn't. The only other possible conclusion was that the anchored ships were moving. Upstream.

Confusion reigned on those ships. All had been anchored with their bows pointing into the current, which would keep the lines taut. But that had been done prior to the storm, when the river had been running south. It had been beyond anyone's wildest imagination that the Mississippi River could possibly reverse its flow. When it did, those anchor lines ran slack, casting that collection of vessels to the mercy of a mighty, turbulent, and confused body of water. The current was inconceivably swift, driven by an elevation difference that far exceeded the gravity head that normally drove the river in these lowlands. Freed by the suddenly slack anchor lines, the ships were swept to the north. When the lines went taut again, the anchors were now behind the ships, and they dragged a considerable distance—miles in a few cases—before they could bite into the bottom and draw the vessels about.

The *Cristobal* was the first to run aground. Fortunately, it was a soft collision with the muddy bottom outside the main channel, and it braked the big freighter to a gentle halt. After confirming that they weren't taking on water and that the cargo was secure from the elements, the captain reassured his nervous crew that they were in no further danger. He kept to himself his own reflections on how lucky they'd been; had the fickle current driven them into the collection of fuel tanks on shore, even with just a glancing blow, right now they'd all be toast.

As he watched the storm from inside the bridge of the mired freighter, two questions in particular piqued the captain's curiosity: (1) the speed of the river (in the wrong direction) and (2) the speed of the wind. In principle, the river current might be measured by timing how long it took a piece of flotsam to drift a known distance relative to the ship. In practice, however, such an observation was impossible in the swirling brew of white-capped waves and foam. In fact, nobody has ever managed to derive a figure for the speed of the river that evening. The captain of the *Cristobal* turned his attention to the slightly more manageable question of the wind speed.

Here in the left front quadrant of the storm, the rain occasionally diminished to the point of allowing some observations. In the beams of the searchlights, the captain could see that the wind was driving the rain horizontally, or essentially from the ship's stern toward its bow. Within that wind was a panoply of debris, including tree branches and sections of roofing. As an airborne limb tumbled past the bridge, the captain punched his stopwatch; he stopped it as the branch leapfrogged the bow rail 350 feet away. Time: 1.5 seconds. He did the arithmetic: 159 miles per hour. Never had he heard of a wind this intense.

A single observation like this would qualify as a gust, not as a sustained wind. But the captain repeated the measurement about a dozen times over a period of several minutes, and the answers differed only slightly. He entered the average figure in his log: winds of 160 miles per hour.

Then, taking stock of the physical implications of a wind that extreme, he scampered down the stairs to ride out the rest of the storm with everyone else below deck.

Around 6:00 p.m., Luke Petrovich got word at the Buras police station that an elderly woman and her adult mentally retarded son were still at their home in Venice, about fifteen miles to the south. Red Lloyd offered to drive, Luke climbed in the front seat, and, with Spiro Pavlovich and the aptly named Camille Gomez sharing the backseat, they headed south. Luke didn't expect hurricane-force winds for another couple of hours.

At first the gusts were intermittent, but as they passed the Boothville Presbyterian church on one side of the road and St. Anthony's Catholic Church on the other, the wind began buffeting the car like a punching bag. They rolled down the windows to try to equalize the pressure, but that didn't help much; the car veered crazily, right and then left. It was clear they weren't going to make it to Venice. They turned around and headed back toward Buras.

It was too late. A fishing tackle store tumbled across the road in front of them, and Red almost rolled the car trying to avoid it. They were back in front of the Presbyterian church again, and that's as far as they'd get. The sanctuary was a single story, but the back section, which contained offices and storage rooms, rose two stories. Remembering that the church was one of the few structures to survive Hurricane Betsy, it seemed like a safe place to take shelter. The men broke in by knocking out a window in the door.

Then, in Luke's words, "The biggest roar I ever heard came through and knocked off the whole back two stories." The four men gazed dumfounded past a stairwell dangling from a couple of remaining rafters. Beyond, the lawn was submerged under swirling water.

When you grow up around the river, you can tell river water; it's not clean like rainwater but has all sorts of scum and bits of flotsam on its surface. Luke tasted it, and to his surprise it was salty. The whole damn river was flowing backward from the Gulf, and it had either broken through the levee or had poured over it.

Louisiana has never been big on zoning ordinances. One finds antebellum plantation homes within view of oil refineries, upscale housing developments next to gas wells, and school buildings adjacent to pipe yards. Fortuitously, that church in Boothville happened to be next to a sewerage treatment plant.

Spiro, shouting that they needed to get to the nearest reinforced concrete tank, took off running through the shin-deep water. The other three watched in horror as a piece of roofing tin came spinning through the air straight toward him. Luke stood transfixed thinking, "My God, it's gonna decapitate him." At the last instant the tin veered off just inches over his head and hit the water, skipping along a couple of times before it disappeared. Spiro made it to the tank's steel stair and began to climb.

Luke glanced around the wrecked church, its remaining walls and roof wobbling in the wind and threatening to collapse at any moment. In one corner, illuminated by a fusillade of lightning flashes, he saw a long yellow electrical power cord. He, Red, and Camille Gomez tied themselves together by putting the cord through their belt loops and then took off through the driving rain and floodwater toward the treatment plant. They heard the church's timbers creaking behind them, but they didn't bother to look back. Seconds later, they heard nothing else but the howling wind.

Red got to the sump tank's stair, grabbed the rail, and pulled the other men toward him. By the time Camille Gomez got that far, he was already waist deep in water. In less than a minute, the flood had risen several feet.

They climbed to the top, seventeen feet above the ground. Although that put them above the water, they had traded that hazard for another: a fierce wind that pummeled them with debris and threatened to sweep them off. Although they crouched shoulder to shoulder, the roar was so loud they could barely hear each other's shouts. Someone noticed a high pressure hose on top of the tank, and they wrapped it around themselves and the railing. Luke thought to himself, "If this is happening to us, what's happening to the rest of the people in the parish?"

Camille Gomez peered down into the sump tank, where sewage was transferred from the main lines into the treatment tanks. The odor was stifling, but at least the tank was empty except for a little bit of slime at the bottom. He climbed inside. His three companions still weren't desperate enough to resort to taking shelter in a sewerage tank.

A windswept sheet of plywood crashed into the steel banister. Had it hit them, it would have inflicted serious bodily harm. Instead, the wind pinned it against the railing and Luke, Red, and Spiro hunkered down behind it. There they huddled for more than an hour, at least partially shielded from the horizontally driven rain and airborne projectiles, but feeling as though they were inside a drum at a rock concert. Luke started wondering which was the biggest hazard—the flood, the wind, the airborne projectiles, or the prospect of a ship or a barge washing over the levee and crushing them all.

As they watched, the floodwaters rose to ten feet, twelve feet, then sixteen feet—just a foot shy of the steel gridiron they were crouched on. In the din of the debris hammering the plywood, there was no way to alert Gomez, inside the tank, about the status of the rising water. Just another foot or so and the floodwater would pour inside, and Gomez would probably drown. But then if the water rose that high, the rest of them would most likely drown too.

Something, possibly a big log, kept hitting the side of the tank—the reverberating booms rivaling the noise of the hurricane. Luke shouted, "Spiro, we're gonna go down any minute." Spiro responded that the tank would hold; he was sure of this because he'd poured the concrete himself. Luke began to untie his rubber boots. "You're not gonna leave," Spiro yelled, grabbing him. "I'm not gonna let you leave."

Luke shouted back that he wasn't planning to go anywhere; he was loosening his boots so when the flood got higher, he could get them off quickly and they wouldn't weigh him down.

With no decrease in intensity, the wind began to shift direction. The plywood sheet started to flap and then fluttered away, exposing them to the elements once again. Luke realized that the eye was going to miss them and that there would be no respite—even temporarily—until this whole hurricane was over.

It got cold, bone-chilling cold—something that just doesn't happen in August in Louisiana. But the water was receding slightly. In fact, unbeknownst to them, a small river of it was pouring out into the swamps through a breach in the parish's back levee. Luke crawled to the tank's interior ladder, poked his head inside, and shouted down to see if Gomez was all right. Gomez shouted back that the smell wasn't great but they'd get used to it. Hoping it would be warmer inside the sewerage tank, Luke swallowed, took a deep breath, and climbed down. Red and Spiro followed.

Their clothing soaked, all four were shivering uncontrollably. They huddled together, but that wasn't enough. To make better use of their body heat, they paired off and took turns lying on top of one another. Neither the bottom nor the top position was particularly comfortable, but it did help warm them a little. They continued this for four or five hours, until daybreak. Luke would later comment, "You've never seen four men who weren't homosexuals get so close to one another."

The flood came in two stages. First, the river sloshed over the levee and inundated everything south of Buras to a depth of five feet. Then, a short while later, a second surge raised the flood to a depth of twelve to sixteen feet.

With that, the Jurjevich home disintegrated. The floodwaters ripped the second floor off the concrete walls of the ground story, leaving those hefty lag bolts securing nothing but the splinters of the sill plates. Leander Jurjevich's wife, his parents, and his two young children climbed into the windowless attic with Charmain's two kittens. Lightning flashing through the gable vents cast eerie moving shadows on the rafters while the rain streamed in first at one end and then the other. They were afloat and spinning in a giant whirlpool. The howling wind muffled any prospect of verbal communication. They huddled, shivering as much from fear as from the sudden cold.

The giant maelstrom of swirling water and debris scoured out the back levee that separated Boothville from the marshlands to the west. The floodwaters gushed through that broken dike, eventually to return to the Gulf of Mexico. The Jurjeviches' attic continued to ride that churning flood for several miles, bobbing, pitching, and spinning in the wind and the current.

Pass Christian, Mississippi, Evening, August 17, 1969

There is never an exact time when a hurricane strikes; there are only times when people become aware that they're experiencing the real thing. And that datum is more a matter of local circumstance and subjective judgment than a matter of objective observation. So it was along the Mississippi Gulf Coast.

It was going on 8:00 p.m. when Ben and his exhausted neighbors took a breather to watch the waters of the Gulf from the second-floor exterior walkway of the Richelieu. For the past eleven hours or so, they had been preparing nonstop—boarding up windows, buying provisions, shuttling cars, and moving furniture to the upper floors. Now the elderly Matthews—she confined to a wheelchair—were secure in a third-floor apartment with a pot roast on the stove, and it was about time to go up and join them.

A police officer scampered up the stairs, tersely greeted them, and then advised everyone to evacuate immediately. When the group demurred, the patrolman pulled out a notepad, made a show of jotting down everyone's name to notify next of kin, and then returned to his car and drove off. His list of names never did make it into any official file.

Ben gazed out at the Gulf and commented that the horizon seemed to be rising. Perhaps he was seeing the storm surge sweeping in, perhaps not. The horizon of a calm sea is always precisely at eye level, and surges indeed rise above that, but was it possible to visually detect a storm surge when it was still several miles offshore? Maybe so, maybe not.

Sheltered from the rain by the overhanging third floor, they dallied awhile to watch the oncoming storm. The beach began to shrink. Channel markers seemed to be disappearing. When the surf swallowed up the last of the sand, there could be no question about the reality of the rising water. Mere moments later, the waves were crashing against the seawall and catapulting plumes of sea spray high in the air. Another few minutes, and Highway 90 was inundated. Now the raging surf was climbing the Richelieu's front lawn.

The complex's three buildings were arranged in a "U" shape, with the opening facing the Gulf. Spanning that gap was a low brick wall that served as a security and privacy fence for the swimming pool. The wall was no match for the onrushing surge. As the bricks tumbled, the waves flooded the pool and crashed against the walls of the first-floor apartments.

A torrent of rain blew into the group's second-floor vantage point. Ben and the others headed upstairs, where the mansard roof enclosed a walkway that encircled the walls of the third-floor apartment entrances. Through the windows in that sloping roof, they could still get a view of the storm—at least for a little while, until those windows began to shatter.

Much of the land on the peninsula lies below the level of the river, and some of it is even below sea level. *(Photo by the authors.)*

The remains of St. Anthony's Church, teetering on a levee. This was Luke Petrovich's first sight on emerging from the sewerage tank that had saved his life. *(USACE, New Orleans District.)*

Top: Many neighborhoods in southern Plaquemines Parish were completely scoured from the landscape. *Left:* Flooding inundated the region between the river levee *(right)* and the back levee *(left)* for a distance of more than twenty miles. *(Photographs courtesy of USACE, New Orleans District.)*

Top: Tractor-driven irrigation pumps removing the last of the floodwaters.
Bottom: With no appropriate buildings having survived on land, the parish set up its emergency headquarters in an old riverboat. *(Photographs courtsey of USACE, New Orleans District.)*

The Richelieu Apartments in Pass Christian, the site of the alleged "hurricane party." The three buildings of the complex were arranged in a "U" shape around the pool. *(Courtesy Ben Duckworth.)*

The remains of the Richelieu Apartments after Camille. *(Photo by Fennell.)*

Hubert Duckworth *(left)* and the Reverend Bill Duncan at the ruins of the Richelieu, moments after they were informed that Ben Duckworth was alive. *(Courtesy Ben Duckworth.)*

Searching for bodies in Mississippi. Dozens of victims would never be found, in all likelihood having been washed out to sea. *(U.S. Navy.)*

Freighters beached at Gulfport. Two of these three large ships had to be cut up for scrap; the third was repaired and refloated. *(Courtesy USACE, Mobile District.)*

Top: The Pirate House in Waveland—built in 1802 and reputedly once owned by Jean Lafitte—as it appeared before Camille. *Bottom:* The Pirate House, after Camille. *(Photographs courtesy Bob Hubbard and McCain Library and Archives, University of Southern Mississippi.)*

Top: Trinity Episcopal Church in Pass Christian, built in 1847, as it appeared before the storm. *Bottom:* Trinity Episcopal Church after the storm. Sexton Paul Williams lost thirteen members of his family in the building's collapse. *(Photos by Fennell.)*

The main street in Pass Christian. (*Courtesy Fred Hutchings and McCain Library and Archives, University of Southern Mississippi.*)

The evacuation of Pass Christian after the storm. Blacks and whites were transported to Jackson and to Camp Shelby in separate buses. *(U.S. Navy.)*

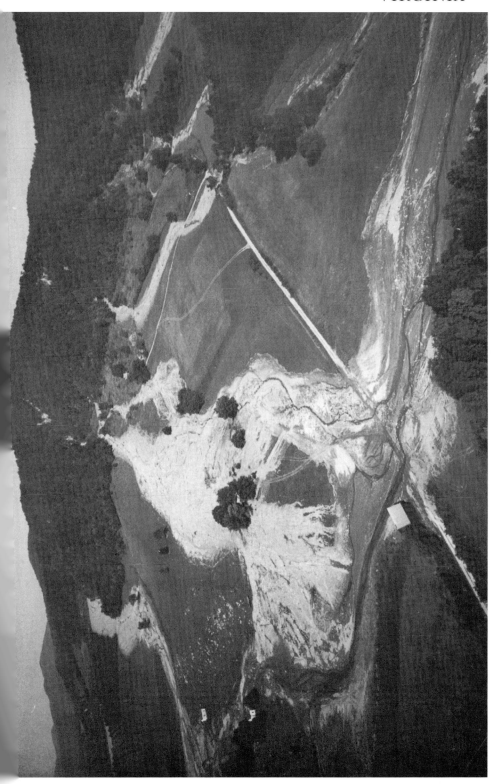

The scarred flanks of Woods Mountain. (*Division of Mineral Resources, Commonwealth of Virginia.*)

Cliff Wood speaking to unidentified helicopter pilots. The command center, consisting of a school bus and a Chevy van with a tarp stretched between the two, is in the background. It was set up on the Highway 29 bypass around Lovingston, which served as a makeshift landing strip for aircraft. *(Courtesy Cliff Wood.)*

Left: An exhausted Sheriff Bill Whitehead in a rescue helicopter. *(Courtesy estate of Bill Whitehead.)*

Opposite: Carl Jr. *(left)* and Warren Raines in the kitchen of their home in Massies Mill. The flood claimed the lives of their mother, father, and three siblings. The high water mark is on the wall behind them. A short distance downstream, the water was much higher. *(Courtesy Eugene Ramsey.)*

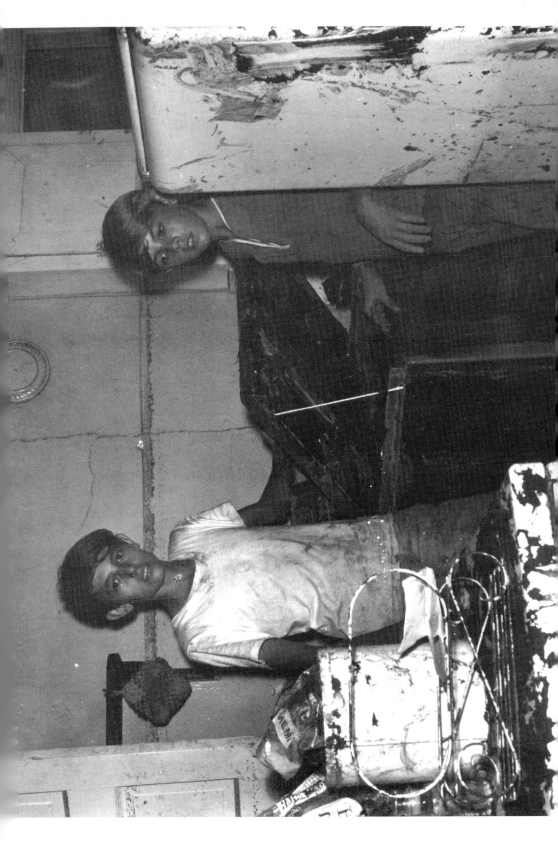

Volunteers searching for bodies near Woods Mill. Even when a body was discovered, it often took a half day to extricate it from the debris. *(Courtesy Russell "Rip" Payne Family and Albemarle-Charlottesville Historical Society.)*

A bridge washout on the Rockfish River. (*Division of Mineral Resources, Commonwealth of Virginia.*)

Two pairs of welded rails were all that remained of the Southern Railway bridge over the Tye River. The bridge was 680 feet long and rose 95 feet above the streambed. This was one of nine washouts along the Southern's rail line. *(USACE, Norfolk District.)*

CHAPTER 9

ANGRY SEAS

Lake Borgne, Louisiana, August 17, 1969

The four sailors on the *Rum Runner* had little choice but to bunk down early. There wasn't enough power to run the lights for very long, and in any case none of them were much in the mood for sitting around talking. Rather than argue further about who was responsible for the navigational blunder that had stranded them short of the pass, they knew what every other man in the world knows instinctively in such a situation—that if they just let the issue die a quiet death, they would all be in friendlier moods the next day. They would then sail leisurely back out into Lake Borgne, tack back in and get it right this time, and laugh about the previous snafu. As far as they knew, Camille was still headed toward Florida or maybe Alabama.

Then the winds began. Weird currents swirled around them, and the *Rum Runner* dragged anchor until it ran aground. The boat banged the shoreline with a series of dull thuds, its wood ribs creaking, caulking working loose from the seams. They started taking on water. They bailed but to no avail. They transferred some of their personal gear to the dinghy, jumped into the now shallow water, and dragged it ashore. They considered crowding into that tiny launch and heading for the pass under the power of the small outboard motor. But the wind was intensifying, and the rain began pelting them like buckshot. Illuminated by the lightning, a fierce chop was kicking up. They knew they wouldn't have a chance of making it to the pass in the overloaded dinghy, so they returned to the schooner.

Forget the bailing; the hull was disintegrating. But it was a wooden boat, and even if it broke up, the pieces would be buoyant. They unrolled a tarp and pulled it over their heads to shelter them on deck and strung a lifeline between the queen's mast and the mainmast for everyone to hang onto. There they sat, side by side, as the boat rocked wildly in the wind and the surf. They made a solemn pact among them. If anyone should get tossed into the water, he was on his own.

Breathing became a matter of timing, of sucking up gulps of air between the dousings from the breakers. They clung to one another, physically close yet each knowing he could do nothing to aid anyone's survival but his own. Under the relentless battering, the sounds of the *Rum Runner's* bursting seams joined a symphony with the roaring wind, the crashing waves, and the booming thunder. They hoped for the eye of the storm—a few minutes of respite to catch their breath—but it did not come. Camille's eye passed just east of them; in sailing from Biloxi to Alligator Point, they had crossed the hurricane's future path mere hours earlier.

They watched through pinholes in the tarp. And what they witnessed next was weird. Despite the angry waves, the waters of the lagoon began receding. If there was any doubt about that, it was dispelled when the *Rum Runner* keeled over ninety degrees and dumped them from its deck into the muck. Lake Borgne, a body of water covering some 240 square miles, had surged temporarily and had essentially run dry.

Twenty miles across the marshlands to the south, the Isleno shrimpers at Delacroix Island were experiencing the same strange phenomenon: that of receding rather than rising water. Whereas Hurricane Betsy, four years earlier, had dealt a devastating flood there, Camille was doing just the opposite: she was drawing water *out* of the region. The Islenos who had remained behind to play out their mooring lines and control their boats during the expected storm surge now found their vessels stuck in the mud with no water more than puddle deep remaining anywhere within sight.

Where did all of that water go? It was driven by the winds and the barometric pressure gradient into the right front quadrant of the hurricane, and when the Mississippi coastline blocked that vicious current's further movement north, it piled up on that shore in a giant storm swell, bearing huge crashing waves on its surface.

Biloxi, twenty-five miles east of Camille's landfall, occupies a six-mile peninsula that varies in width from a half mile to two miles, with the Gulf on the south and a back bay at the north. There, at St. Michael's Catholic Church, young Father George Murphy set two chairs out on the rectory veranda, figuring that he and the pastor would spend an interesting evening watching the storm. Recently arrived in the United States from his native Ireland, Murphy wasn't being foolish so much as inquisitive; he wanted to learn everything he could about this place called the United States, and hurricanes were part of that agenda. He went to his bedroom, changed into his pajamas and slippers, and then invited the pastor to join him as he stepped back outside.

The chairs had vanished. Buffeted by a howling wind and pelted by rain, he stood in awe at the sight of the debris-filled sky and the breakers crashing onto the highway. "So this is what they call a hurricane," he whispered to himself. "Very impressive." He broke his trance only when Pastor Cavanaugh dashed out of the house, tossed him a blanket, and shouted to him to get to the church. It was wise advice; soon nothing would remain of the parish house but matchwood.

The power was already out. As the two priests lit candles in the sanctuary, Murphy noticed a bowling pin floating up the main aisle. He recalls thinking, "If that bowling pin is in the church, what happened to the bowling alley?"

The church windows began to shatter. The water rose farther, and they climbed the statues of a side altar, which some builder had had the foresight to anchor in place with rebar. With the floodwaters lapping at their feet and the building shuddering while its interior swarmed with airborne miscellanea, the candles flickered out and Fathers Murphy and Cavanaugh were pitched into darkness.

All in all, it was not a good night to be in a church along the coast. Just a few blocks down the road, Biloxi's Episcopal Church of the Redeemer was collapsing under the winds and floodwaters. Today, all that remains of that once-impressive place of worship is the bell tower, with a memorial to Camille's victims now occupying the former site of the church itself. Farther to the west, not a trace remains of dozens of churches that stood there before Camille put them to a test they would fail.

The churches along this coast had been designed with a mind

toward aesthetics, economy, and minimizing the discomfort of the oppressive Gulf Coast summer heat and humidity. None of the architects, however, had given much thought to how these buildings would fare in high winds; for the most part, their frames were held together with nails rather than bolts and steel straps, and none but the newest of them depended on anything other than gravity to secure them to their foundations. As their broad walls and high roofs deflected Camille's hefty winds, Newton's law of action and reaction went to work, and the structures began to sway, twist, and contort, pulling loose one fastener after another. As rafters separated from walls and studs pulled away from sills, the previously staid houses of worship morphed into weird giants doing devil dances in the wind.

Those unfortunate enough to have taken shelter in these normally peaceful places heard the shrieks of the wind punctuated by the creaking of loosening nails. Art and religious artifacts tumbled from the walls as plaster plummeted from the ceilings. Hanging light fixtures swung crazily, illuminated stroboscopically by the lightning. As flying debris smashed some of the windows, others blew in when their frames flexed loose. Crouching under the pews offered protection for only a short while before the storm surge struck.

St. Thomas Catholic Church in Long Beach was built more substantially than most of the other churches, with stout timber truss work cross-bracing its roof. It was one of the few that did not shift off its foundation. It was destroyed, nonetheless, by the battering of two barges that floated in on the floodwaters.

In Pass Christian stood Trinity Episcopal Church, an architectural gem dating from 1849. When the winds began to howl, sexton Paul Williams, his wife, his son-in-law, and his eleven children and two grandchildren left Williams's brick home for the presumed better safety of the old church. Mere longevity, unfortunately, is no proof of a building's invulnerability to the forces of nature, and the church complex unfortunately stood on lower ground than Williams's own home. The Williams family had barely gotten inside when the storm surge struck. With the Gulf waters pouring into the sanctuary, it was too late to reconsider their decision. The adults began hoisting the children into the rafters. Unbeknownst to them at the time, the rectory next door had already collapsed, and the minister's wife was drowning.

Before Williams could follow the children into the rafters, the building started swaying, and he heard a loud cracking sound. The failure was complete within seconds. The next thing the sexton knew, he was bobbing to the surface a few feet from an uprooted floating tree. Somehow, the surge had swept him, his eldest son, and his son-in-law out from under the collapsing church. Not so fortunate were his wife, his other ten children or his two grandchildren, none of whom would ever be seen alive again.

At Keesler Air Force Base in Biloxi, the wind was howling by 9:00 p.m. Greg Durrschmidt and a hundred other young airmen were confined to quarters in a three-story concrete dormitory ten feet above sea level and three blocks inland. None of them had been given any instructions other than to remain where they were.

The power failed. Debris peppered the building in what sounded like machine-gun fire. The ventilator covers ripped away, followed by crashing noises as they careened off parked cars. The upper-story windows started to shatter. The men from the top two floors ran downstairs, dragging their mattresses. Portable radios picked up nothing but static; they were completely cut off from outside communication. The roof failed and water cascaded down the stairwells, joining the streams that were flowing in under the exterior doors. Sitting in the rising water, the young men pulled the mattresses over their heads, wondering whether the greater threat was from the building coming down or the flood coming up.

Unbeknownst to them, they were only on the eastern fringe of the storm, with winds of 81 miles per hour gusting to 129. The situation was far worse closer to the storm track.

One of the last men to evacuate a Transworld Drilling Company rig that morning had the foresight to switch the wind speed recorder to double-scale and to leave it running on battery power. That offshore instrument, some thirty miles east of Camille's storm track, recorded 172 miles per hour before it jammed. Several wind gauges in the eastern sections of St. Tammany and Washington Parishes, Louisiana, registered 160 miles per hour—the same figure that was reported by the captain of the *Cristobal* at Pilot Town. These measurements were all made between 30 and 50 miles from the eyewall, and several were at a considerable distance inland. No wind speed instruments near the path of the eyewall survived. In fact, not much of anything survived there.

Human curiosity being what it is, many people who are interested in superlatives such as the "highest wind speed" are disappointed when they learn that there is not a clear answer. A hurricane is not a steady-state phenomenon; it is a dynamic fluctuating vortex of wind and rain that does not have a single speed. For purposes of standardized comparison, the National Hurricane Center tries to report the highest average speed measured over a one-minute time period; any shorter interval than that is called a "gust." Ideally, the anemometer should be at a height of thirty feet above the ground (actually, ten meters); in practice, however, this condition is not always easy to meet. And anemometers, which are mechanical devices that need to catch the wind in order to measure it, have their mechanical limits. Just as a bathroom scale is unlikely to survive being stomped on by an elephant, anemometers are notorious for disintegrating when exposed to hurricanes.

Barometric pressure, on the other hand, can be measured easily and reliably and indoors. By 1969, the relationship between barometric pressure and the altitudinal variation of wind speed was fairly well understood. Using the reconnaissance flight-level wind data and the known barometric pressure at ground level at the time, meteorologists retrospectively computed that the sustained sea-level wind speed at Camille's eyewall just before landfall reached 201 miles per hour. No hurricane before or since has had an objectively established wind speed anywhere near this high. Still, the figure that appeared in all of the reports was the more conservative 172 miles per hour—the value that was actually measured with an instrument at ground level. Even this lower figure set an intensity record.

High winds generate several types of forces on a structure. The most obvious is the inertial force, which arises when a building stops (or alters) the forward momentum of the moving air mass. This is the force we feel on an outstretched palm stuck out the window of a moving car. Clearly, this inertial force is greatest when the palm is perpendicular to the airstream—the orientation where it presents the maximum frontal area.

It might be tempting to assume that a doubling of the wind speed will double the inertial force, but Mother Nature has chosen a different formula. Each doubling of the wind speed actually generates four times the inertial force; each tripling of the wind speed generates nine times the force; and if the wind velocity increases by a factor of ten, the inertial force increases by a factor of one hundred. Camille's winds generated a force of several hundred pounds on each square foot of obstructive surface—which added up to dozens of tons for some of the larger buildings.

These numbers relate only to the inertial force of the wind. There are also aerodynamic forces, which are highly sensitive to a structure's geometry. Hold a limp strip of paper in front of your mouth and blow over the upper surface, and the paper will rise into the airstream. Similarly, roofs and other man-made structures have a tendency to be lifted and carried off by high winds.

Then there is the prospect of tornadoes. These tightly wound vortices arise when adjacent bands of the hurricane happen to acquire different speeds. The effect is easily demonstrated by holding a pencil between the two facing palms of your hands, then slowly moving one palm over the other. The rapidly spinning pencil is like a pocket of spinning air. This tornadic effect can occur even hundreds of miles from the hurricane's eye.

The landfall of a severe hurricane is a busy place. The wind batters buildings into splinters, sucks the rubble up into the storm, and flings the debris at high speed into the windows and walls of other structures. It generates tornadoes that snap off utility poles and uproot trees. Its right front quadrant pushes a surge of sea ahead of it. Flood currents sweep foundations from under the homes the winds are smashing. And everything about the event is noisy.

So it was with Camille. And on a scale that dwarfed any hurricane ever experienced before on American soil.

Pass Christian, Mississippi, Nightfall, August 17, 1969

At the Richelieu Apartments, it was around 10:00 p.m. when the power went dead. Ben Duckworth and seven others, including the elderly Matthews couple and Rick and Luane Keller, had just finished eating the pot roast in the borrowed third-floor apartment. In another apartment, manager Merv Jones and his wife were hosting a handful of other tenants who had decided to stay through the storm. On the second floor, Mary Ann Gerlach and her husband, Fritz, were still napping. In total, about twenty-five people remained in the complex.

Ben and his companions began to hear crashes and thumps from below. Around 11:00 p.m., the entire building began shuddering and rocking. Sections of plasterboard peeled loose from the ceiling.

Ben ventured out into the hall. Most of the windows in the mansard roof had been shattered, and the wind howled through the corridor. He shielded his head, made his way along the inner wall to the main stair, and aimed his flashlight down the stairwell. He blinked and rubbed his eyes and then swept the light back and forth. Unbelievably, the water was just a foot shy of the third floor. Anyone who had stayed on any of the lower levels had either drowned by now or was struggling for life out in the flood.

This was serious. Ben staggered back to the apartment and bolted the door behind him. Everyone fell silent. His first impulse was to blurt something like "We may be in trouble," but when he saw Zoe Matthews's sad eyes pleading for a reassuring report, he caught himself. The last news they'd heard—or thought they'd heard—just before the power outage was that Camille's eye would hit the coast at 11:30. Ben glanced at his Timex: it was 11:16. "If we can hold out for fourteen more minutes," he said, "we've got it made." With that, Ben patted Zoe's shoulder and wheeled her under a doorframe.

Mere moments later, they heard the ominous screeching of nails pulling loose from studs and joists. The walls and ceiling split open. The room, and in fact the entire building, began twisting and swaying crazily. Rainwater poured in through the ruptured roof.

Ed Bielan suggested they jump out the windows and take their chances in the flood. Ben pointed to the breach in the roof. The way to go was up.

Hopping on Ed's shoulders, Ben grabbed the ragged sides of the opening and poked his head above the roofline. The condensing unit of an air conditioner partially shielded him from the driving wind and rain. The lightning was nearly continuous, yet the visibility was a mere few feet. All those flashes illuminated little more than a great blur of horizontal rain.

At five feet seven inches, Ben was not a big man. He was, however, muscular—the result of a rigorous exercise regimen he'd begun in his teens as an alternative to enduring a surgical correction for several malformed vertebrae. Although the draft board hadn't been willing to take a chance on his fitness for military duty, he was in pretty good physical condition. He hoisted himself out onto the roof, planted himself at the edge of the opening, and extended his outstretched hands inside. With some pushing from below and Ben pulling from above, the group began evacuating the doomed apartment. Ben did not pay much attention to what happened to those people after they got on the roof; his focus was on helping them get out. Rick Keller struggled out with Luane on his back, and in an instant—swish!—they disappeared. Ed Bielan climbed out with Mrs. Matthews on his back, and they too were gone. In fact, as soon as each of the group moved beyond the partial protection of the air conditioner unit, they were blasted by the full force of a wind that was clocking in around 170 miles per hour. Nothing short of a lineman's harness—and maybe not even that—could possibly have tethered those men and women to that roof under those extreme conditions.

The building wobbled crazily as Ben pulled several others outside, including a woman whose name he never knew. Near one corner of the disintegrating Richelieu, he caught a glimpse of a creepy sight—a pale orange glow in the shape of a huge vertical cylinder. Simultaneously, he heard a fusillade of loud cracking sounds from below. Candles and plates of food skidded off the counters and onto the flooded apartment floor. He beckoned to Mr. Matthews. The old man shook his head and backed away into the shadows of the far corner of the room, his palms outstretched as if to fend off some grave threat. That was the last image Ben had of him.

The next thing Ben was aware of, he was bobbing in the water, gasping for breath, barefoot, his glasses gone. The three-story apartment building had completely disappeared.

Prior to an imminent catastrophe, there is always a sense of bonding, a heightened sense of community, even with strangers. In the face of imminent death, however, there is always that very personal struggle to keep oneself alive. When it succeeds, such a survivor often limps away bruised emotionally for years. He wonders why he of all people survived, why others did not, and whether his personal survival, and the decisions that led up to it, may have somehow contributed to the deaths of others. The question of whether Ben had done all he could to help the rest of the group would haunt him later, over and again. For now, though, his main order of business was to try to keep himself alive.

Finding something to clutch onto was the easy part; the water was blanketed with splintered boards and all manner of other floating debris. Most of the flotsam, unfortunately, was not very big and not very buoyant. Worse, most of it was riddled with nails and other injurious protrusions. The rain felt like a continuous barrage from a BB gun. His head barely above water, Ben struggled to keep himself facing downwind as he scooped up as much debris as he could circle his arms around. The nails shredded what was left of his clothing. The unmerciful waves jerked one piece, then another, of his makeshift life buoy from his grip.

He let go and tried to swim. All that accomplished was to tire him and to destroy the last vestiges of his shirt. He looked right, then left. A flash of lightning illuminated a ghostlike mass. It was the top of a tree, either uprooted or snapped off by the storm and now floating horizontally. He fought through the branches and grabbed onto the trunk. He began climbing onto it, but that didn't work. Just a mere foot or so above the water, the wind was so severe that it knocked him off. Then the trunk began rolling. Not very fast, but fast enough that there was no way for him to rest. As he repeatedly grasped for new handholds, the bark scraped his arms and chest raw.

When you're floating in a current, you have no sense of the water's movement. The surface waves, yes, you certainly feel them. But where this great flood was headed, Ben had no idea. For what seemed like hours, he struggled to keep his hold on that floating tree, fervently hoping that he wasn't drifting out to sea. Then he felt an abrupt thud, and his tired arms lost their grip. The tree, blown by the wind, drifted out of his reach. This was the end, he figured; this is where he dies.

The thud, however, had come from a collision with something, and that something turned out to be a big solid live oak that remained rooted in the soil beneath the floodwaters. It was close enough that the next lightning flash revealed it even through the sheets of driving rain. Ben took a couple of strokes, caught one of the branches, and climbed out of the water. His only remaining garment was his tattered swim trunks. The rain pelted his bare flesh, adding insult to the hundreds of cuts, scratches, and punctures that covered his body.

The wind was so intense that it sucked the breath right out of his lungs. He tried to face into the storm and get a big gulp of air that way, but that didn't work either; the blasting rain threatened to beat out his eyes right through his closed eyelids, and opening his mouth in that direction drove bullets of rainwater straight down his throat. Coughing and gasping, he shinnied his battered body to the main trunk.

Like most live oaks, this one had a deeply corrugated trunk. Ben pressed his face into one of those cleavages and found that he could breathe a little. The relentless pelting continued on his bare back, his thighs, his calves, and his feet, but at least he could breathe. It grew cold, deathly cold, and he shivered.

But at least he could breathe. And as long as you're still breathing, you have a chance of making it.

It was the highest storm surge ever experienced on a U.S. coast. The Army Corps of Engineers, in its follow-up studies, would report a "reliable" high-water reading at Pass Christian of 22.6 feet above sea level and "less-reliable" figures of 24.2 and 24.6 feet. For a reading to be "reliable," however, the corps insisted that it be based on a *still-water* mark measured inside a surviving building. The "less-reliable" figures were measured inside buildings whose roofs had failed and were therefore at least partially exposed to the elements. Given that few buildings are capable of even partially surviving a storm surge of such monstrous power, it seems unlikely that this small sample of three readings tells the full story of the severity of the flood.

There is reason to believe that, at least at the site of the Richelieu, the storm surge topped out at about 28 feet above sea level. Although it is no longer possible to verify such a figure definitively, the reasoning runs as follows: The ground-floor slab of the Richelieu was approximately 11 feet above sea level, the second floor began 9 feet above that, and the third floor was another 9 feet higher. That put the third floor at an elevation of about 29 feet above sea level. Ben Duckworth reported that, shortly before the collapse, the water in the stairwell was within one foot (about a step and a half) of the third floor. If the structure had survived, this still-water mark would have remained to be confirmed by the corps's investigators, and it would have apparently been about 28 feet above sea level.

But neither the Richelieu nor any other structure in its immediate vicinity survived. Thus, the figures usually quoted for Camille's storm surge are around 24 feet, although some sources do squeak this upward a bit.

Regardless of the precise figure, the evidence is consistent within a couple of feet. Clearly, a great deal of seawater piled up on the Mississippi Gulf Coast that terrible night of August 17, 1969. Some of this water was sucked in from the shallow lagoons and swamps to the west, but most of it was driven in from the Gulf. The currents were horrendous.

At Gulfport, three deepwater freighters were moored at the banana wharfs: the World War II victory ships *Alamo Victory* and *Silver Hawk* and the only slightly younger *Hulda*. These were sturdy but slow ships, having top speeds of around fifteen knots, which rendered them incapable of outrunning Camille. All three had international crews.

Under normal circumstances, language differences do not seriously hamper the functioning of an experienced team of sailors. The men know their jobs, and there are always those who are multilingual enough to pass important information through the ranks. Even in emergencies at sea, everyone is quickly informed of the essence of the problem and the possible outcomes. At the Gulfport docks, however, it was beyond any of the men's imagination that their eight- to ten-thousand-ton freighters berthed in a relatively sheltered sound might be lifted by an angry sea and deposited on dry land.

There would be arguments and counterclaims about the actual sequence of events in the triple shipwreck. There would be complaints and, later, lawsuits by seamen who felt they hadn't been properly advised about the meteorological hazard. There would be liability disputes about which ship broke loose first. Possibly it was the piers that failed first and cast those three ships adrift on the storm tide.

A freighter, even when moving slowly, packs an enormous amount of momentum that will crush virtually any obstacle it might collide with. A docked freighter, when lifted by rising water, also has a huge amount of buoyancy that will snap mooring lines or yank out piers. It isn't clear whether all three captains were on their ship's bridges at the time the storm surge struck, but no matter, for if they were, they couldn't possibly have seen or done much of anything anyway. None of the crews were likely to have seen anything either, for it was impossible to venture on deck in the driving wind and there were virtually no lights on shore.

Adrift on the storm surge, the three ships collided with each other multiple times while they pulverized several of Gulfport's dock facilities. The *Hulda* and the *Silver Hawk* took the worst of the beatings. Sunrise would reveal these two freighters marooned upright on the beach, side by side, their hulls gashed, their machinery ruined by intruding saltwater, their bewildered crews unable to get to land without climbing down precariously swinging forty-five-foot rope ladders. It would prove uneconomical to repair those two ships; they would be cut up for scrap at the site. The *Alamo Victory* fared slightly better, apparently because one of its boilers had been kept fired during the storm. Although it also was grounded and dented up, its hull remained intact, and it would ultimately be refloated.

Although there were no deepwater facilities along the section of coast west of Gulfport, dozens of shrimp boats, barges, tugs, and pleasure craft were swept inland and deposited in the debris piles that had been the beachfront properties of Long Beach and Pass Christian. A large oil storage tank came to rest near the highway after floating several miles from its original position. All buoys and navigational aids were gone. And the furious storm tide tore off the rails and every single cross-tie from a stout concrete railroad bridge spanning the Bay of St. Louis.

Assuming adequate visibility, what would people have seen that terrible night? Waves breaking toward shore, certainly. But that particular evening, the waves behaved weirdly—they crashed in without ebbing out again. One after another, each breaker piled up on the runup of the one before it, building the water level higher and higher and sweeping ever deeper into the residential areas and coastal lowlands. This flooding was not simply a matter of rising water; the overall effect added up to a fierce incoming current that swept buildings and storage tanks off their foundations, uprooted trees, beached ships and boats, and claimed hundreds of lives.

CHAPTER 10

DAWN

Pass Christian, Monday, August 18, 1969

Around 4:30 a.m., with the wind diminished to a stiff breeze, Ben saw strange flashes of light below him. He thought he was dreaming before he realized it was a group of men who wielded flashlights, beckoning him down from the tree. His nearly naked body was peppered with cuts and bruises, one leg sporting a nasty gash. His wristwatch and his college ring were gone. It took a few moments for his sluggish brain to register that the flood had receded.

His aching muscles trembling as much from the cold as from the ordeal, he started to shinny down the slime-covered trunk. His tightest grip wasn't enough to keep from accelerating toward the ground. Two brawny arms grabbed him around the pelvis and broke the fall. A pair of muscular hands steadied him on the mushy ground. The big black man turned to his companions. "This child, he cole."

Heaps of debris sparkled in the flashlight beams; everything was littered with shards of broken glass. Someone draped a soggy throw rug over Ben's bare shoulders. Words were exchanged that he didn't quite decipher. He felt himself being hoisted onto someone's shoulders. His mind was too numb to ask where he was being carried.

In the morning twilight, a burly black man trudged up to the white high school with a young white guy slung over his back. He handed Ben over to a relief worker and then left to resume his search for "his people." Nobody bothered to ask his name.

The school was pandemonium. Some victims had terrible injuries, others appeared dead, and a few *were* dead. Frenzied women screamed for lost children, and children cried for lost parents. There were no doctors. It wasn't apparent who among the volunteers were licensed nurses. Frantic survivors continued to arrive as some of the injured left without telling anyone. Nobody was keeping records about anything.

A woman led Ben to the science room, gave him a blanket, told him to lie down on one of the black slab tables, and then left. Ben tried to doze off but couldn't. Somehow, he felt compelled to return to the apartment. Rather than struggle through the throng of humanity in the hall, he climbed out one of the classroom's broken windows.

Not a single building within sight had escaped damage, and the spectacle of devastation only worsened as Ben approached the Gulf. In places, the wreckages of homes, smashed cars, and splintered boats were heaped twenty feet high. Downed trees blocked every road, and as he crawled over and through them, he reopened some of his wounds. His entire sense of place was fractured. The single landmark he did recognize was the concrete seawall—still intact only because it had been submerged so far below the waves that it escaped their scouring action. There he saw a man with a camera, a young boy beside him. "Hey, Dad," the youngster shouted, "look at that guy! Get a picture! Get a picture!"

He trudged west, mechanically stepping over and around the jetsam, his mind still foggy, the cuts on his bare legs and feet oozing fresh blood. He bypassed the site of the Richelieu before he realized he'd gone too far. He turned around, while trying to reestablish his bearings, and bumped into a disheveled blonde woman who was crying uncontrollably. He'd shed a few tears himself when he was up in the tree during the worst of the storm, and at one point he'd even prayed to God to take him, but right now he wasn't much interested in crying or seeing anyone else do so. Numbly, he ignored the woman and staggered on. He spied a cluster of bent and truncated piping jutting eerily into the air from a concrete slab, and then he recognized the swimming pool.

His first impulse was to look for "stuff," but he couldn't decide what stuff or where to start. He noticed someone hobbling around on a makeshift crutch fashioned from a tree branch. He squinted, trying to compensate for his lost glasses. It was Rick Keller. Ben ran over and tried to hug him, but Rick wasn't interested.

Destruction in Pass Christian. The eastern eyewall passed through this region. Not a structure remained standing within four blocks of the water. *(USACE, Mobile District.)*

"Where's Luane?" Rick asked. "Have you seen Luane?"

Ben shook his head.

Rick turned away and limped off to the north, vainly shouting the name of his pretty young bride over and over.

Ben resumed his irrational search through the rubble. What he was looking for he couldn't say. Only when he stepped on a dismembered framing timber that drove a nail deep into his foot did it dawn on him that he'd been trekking around barefoot and practically naked. He inspected the bottoms of his feet. Both were covered with cuts, blood, and grime. Suddenly, his battered body begged him to give it a break. He staggered back to the high school, climbed back in through the same broken window, found the blanket still there, curled up on the lab table, and fell deep asleep.

His roommate, Buddy Jones, had been assigned to one of the amphibious vehicles dispatched to Pass Christian for search and res-

Boats became hazards to homes, and homes were hazards to boats. *(USACE, Mobile District.)*

cue operations. When he glimpsed the site of the Richelieu, he was stunned. It wasn't until mid-afternoon, however, while dropping off a load of victims at the high school, that Buddy finally got a chance to ask about Ben. A frazzled woman volunteer told him that a young man of that approximate description was back in the science lab.

Buddy found Ben curled up around a gooseneck sink faucet surrounded by Bunsen burners, a rumpled blanket under his head, fast asleep. He shook Ben's shoulder. Ben rolled his head and squinted at him. "Hey, Buddy," he said, "I got the apartment all cleaned up for us, man." Then he buried his face back in his makeshift pillow.

Buddy shook Ben awake, helped him outside into the LARC, and had the driver take them to the Gulfport naval hospital. There, he poured a couple of gallons of hydrogen peroxide into a bathtub and told Ben to strip off his shorts and climb in.

If the 170-mile-per-hour pellets of rain had been painful, this was ten times worse. Ben screamed as bubbles fizzed from every square inch of his flesh. A nurse came in, tended to the gash on his leg and the puncture in his foot, and gave him a tetanus shot. Buddy scrounged up some clothing, and the two went next door to crash in the medevac office. Ben immediately dozed off again. The thought of trying to get word to his parents didn't even occur to him.

Lake Borgne, Louisiana, Monday, August 18, 1969

At Alligator Point, the sea returned before daybreak, floating much of the splintered decking of the *Rum Runner* out into Lake Borgne and leaving only the schooner's capsized hull and masts mired on shore. Wrapped in the tattered tarp, the four hapless sailors chased the snakes out of the battered hulk and huddled there for warmth. They had no provisions—no water, no food, no cigarettes. Ronald Durr was still chewing the same stick of gum he'd popped into his mouth the night before, when the storm had begun. Later, he tacked that piece of gum up on a bulletin board in his office, where it would remain for more than twenty-five years. As for cigarettes, he would never light another one from that day on.

The dinghy was brimming with water, and its outboard was ruined. Although there were a few dozen recreational cottages within a couple of miles, there was good reason to believe that those camps hadn't fared any better than had the schooner. Then they heard aircraft—the first of hundreds that would ply the skies over the disaster zone during the coming weeks. They jumped and waved and shouted, as if their shouting could possibly make any difference. One of the men found a piece of broken mirror from the head and began flashing the reflection of the rising sun at the planes. It was that mirror that did it. A seaplane from the J. McDermott oil company circled in and landed on the now-still lagoon, taxied to the shoreline, and throttled back. As the four castaways waded out, the incredulous pilot stepped out onto one of the pontoons and shouted, "What the hell are y'all doin' out here?" Then he offered to fly them, two at a time, to New Orleans.

In a peculiar reversal of the previous night's "each man for himself" pact, however, the four men now found themselves committed to stay together. The pilot radioed their position to the Coast Guard and then took off into the sun, leaving the men with the promise that he'd check back on them in awhile.

A short time later, a military helicopter buzzed in over the horizon. The ground was too mushy to risk landing, but the pilot descended to a few feet and hovered while a young Coast Guardsman in the open hatch flashed a green board. Chalked in block lettering were the words "DO YOU NEED HELP?" "Gimme a break," Durr muttered, his words lost in the din of the whirling blades. "*Do we need help?*"

The crewman dropped a rope ladder. Ragtag, exhausted, and fighting the downwash, the four men boosted and pulled each other into the chopper. Earl Hover sat at the side window and gazed at the battered bones of the *Rum Runner* until they disappeared in the distance, a tear trickling down his cheek.

Lower Plaquemines Parish, Daybreak, August 18, 1969

Commissioner Luke Petrovich climbed out of the holding tank and stared numbly at the brutal scene of destruction. Except for the sewerage plant that had saved him, not a single structure within miles had survived. The debris blanketing the floodwaters, still nine feet deep in places, undulated like the ragged back of a gargantuan sea monster in its death throes. Of St. Anthony's Catholic Church, all that remained intact was the truncated steeple, now aground on the levee and leaning crazily like the Tower of Pisa. An irrational inner voice whispered to him, "Luke, it's all your fault."

As if in response to his agony, the rising sun broke brilliantly behind the steeple's cross. "Is this a message?" he asked himself. "Am I gonna change my ways?"

He heard an outboard motor. A bass boat that had somehow survived on one of the levees was cautiously weaving its way toward him through the flotsam. He waved. The boat's owner recognized him and steered in.

As they motored upstream, they noticed a smell, faint at first then unmistakable—of cooking tomatoes. A workboat was stranded on the levee with its stern in the dry, and Luke yelled, "Hey, anybody in there?"

The hatch flew open and out bolted a swarthy fellow with his cap on sideways, waving a cooking ladle dripping with macaroni. It was One-Punch Hogan, a squirrelly grin on his leathery face. He squinted. "Who dat? Luke? Damn, she blew last night, didn't she?"

From the *Mary Kay,* Leander and Joe Jurjevich scanned the levee for the *Buffalo.* All they saw on that narrow strip of ground was a confused mass of splintered lumber, dented oil tanks, grounded marine equipment, and a smattering of confused cows wandering around the carcasses of their less fortunate bovine cousins. A short distance downstream, they spied a crevasse in the levee where they could paddle from the river side into the parish. They borrowed a pirogue from the crew boat's captain and started paddling.

Overnight, the south end of the parish had become a lagoon some twenty miles long and averaging around a half mile wide. The floodwater had no way to retreat. Barely a home remained, and even most of the larger buildings were gone. In the distance, they saw plumes of black smoke billowing against the otherwise blue morning sky. Several floating roofs were on fire, and burning asphalt shingles generate ungodly heavy smoke. A couple of river barges were aground, one in the transformer yard of the flooded power substation. Animal carcasses, foliage, and mangled furniture floated everywhere. The only reason there weren't many large trees bobbing around is that very few of the big ones had survived Hurricane Betsy, four years earlier.

"It was somewhere around here," Leander finally said. "The house, it was somewhere around here."

Joe was silent. There was no evidence whatsoever of the Jurjevich home nor anyone who had been in it.

They heard an outboard motor. To an outsider, it might seem incongruous that a small motorboat could have survived the violence of the storm, but in fact a handful of them did. Numerous locals were experienced about hurricanes; they knew about the prospect of flooding and the importance of having a boat in the aftermath, and a number of them had secured their craft high up on the levees rather than trailering them out. Despite the various failures in the dikes, a few of those boat owners had been lucky.

Leander and Joe paddled toward the approaching boat. A wet and

Wrecked boats and homes washed up on the back levee at Fort Jackson.
(USACE, New Orleans District.)

haggard-looking Croatian was sitting at the bow. "Luke?" Leander shouted.

Luke gestured to his helmsman to rendezvous. "Leander?" he shouted back. "Jesus, is your family gonna be glad to see *you!*"

Luke related that he'd spotted the five Jurjeviches on their rooftop a mile or so out in the swamp, waving Margarethe's (Leander's wife) white petticoat. A local teacher who had spent the night on the upper floor of the high school went out in his small aluminum fishing boat and brought them in to the weather observatory in Boothville. They were a little shaken up, but everyone was okay.

"I'm sorry about your house," Luke added.

"God's blessed me with what I value most," Leander assured him.

As Leander and his brother paddled toward the weather observatory, Luke continued north to the police station in Buras. Helicopters were already buzzing overhead. The oil companies couldn't afford to dally and neither could Luke.

The towns of Nairn, Empire, Buras, Triumph, Boothville, and Venice lay submerged. The riverbanks were littered with a tangled hodgepodge of once-expensive drilling and marine equipment, rendered essentially worthless overnight. A major consolation was that more than seventeen thousand people had evacuated. The flip side was that all those people were going to want to return real soon, but there was no place for anyone to live—not even a good spot to pitch a tent. Although the parish had recovered after Betsy, what were its chances of recovery after this one? Where do you even start?

The ground floor of the municipal building was still flooded, but upstairs some thoughtful person had brewed fresh coffee over a Sterno stove. Luke took a sip and was on the radio before he was out of his wet clothes. As director of public safety, he was in charge of coordinating the recovery, and things needed to get started. He started giving directives. Route 23 was to be barricaded at the north end of the parish, just south of Belle Chase. Nobody, absolutely nobody, was to be let through unless they had official duties, and that included news reporters. Boats were desperately needed—shallow-draft boats. Make sure, he ordered, that each one brought extra cans of fuel. Owners would be reimbursed later; yes, he guaranteed that. And send food and water for the parish employees and the survivors down here. Before long, small boats began tying up at the flooded front entrance to the police building.

The next radio calls were to the governor and the Army Corps of Engineers. Luke had worked with the National Guard after Betsy, and that had gone pretty well. Now he would request them again. As for the engineers, he desperately needed their help. Yes, technically they were feds, but in reality most members of the New Orleans District of the Army Corps of Engineers were southern boys. Luke was comfortable working with them, and so was the Guard.

The engineering task involved a lot more than simply filling the breaches in the levees. The big and urgent problem was how the heck they were going to get rid of that giant pool of water—around 1.4 trillion cubic feet of it, more than 11 trillion gallons. All of it was at or below the level of the river, and much of it was actually below sea level.

There was a television show, popular in the early 1960s, called *Naked City*. Each episode focused on a fictionalized resident of metropolitan New York and followed that person through hardship and difficulty to his or her ultimate triumph over adversity. The program always ended with the voice-over: "There are ten million stories in the naked city. This has been one of them."

Most of the reporters and journalists who descended on the Gulf Coast in the disaster's aftermath took a *Naked City* approach to their reporting, akin to "There are tens of thousands of stories in the naked hurricane; this is going to be one of them." The published stories—and hundreds of them appeared in local and national newspapers, telecasts, and news magazines—would be suitably gripping and in many cases gut-wrenching, but in total they fell short of creating any cohesive picture of what had happened.

When the emergency power failed at one of the hospitals and threatened the life of a woman who needed an iron lung to breathe, a physician scavenged batteries from parked cars and successfully jury-rigged them to the machine. Another surgeon performed an emergency appendectomy by flashlight. From their residences on higher ground, some people witnessed half-swamped cars floating by, their headlamps beaming eerily under the water. The FBI was called in to secure a large cache of weapons that had been spilled from the flattened home of an elderly woman and her reclusive son. It was as if those tens of thousands of surviving victims had each experienced their adventures in a vacuum, independent of what was simultaneously happening to uncountable other victims, disconnected from meteorological science, and isolated from the social and political contexts of the times.

In the early hours of the flooding, nearly a thousand people were rescued by LARCs from an amphibious transportation company that had been conducting a routine drill that Sunday and that didn't bother to wait for orders from anyone before springing into action.

Later, dozens of folks had to be rescued from a couple of LARCs when they were damaged in collisions with submerged objects. As these and other stories confirmed the broad scope of the human impacts, various gaps also became apparent.

A major disconnect related to Plaquemines Parish. Because that peninsula had been sealed off after the disaster, almost no news was reported from that region for several weeks, and even today many people who think they know about Camille don't realize the extent of the devastation in southeastern Louisiana. There was, however, another kind of glaring omission from the contemporary accounts. With but a handful of exceptions, they ignored about one-third of the affected population: the blacks. Many readers were left with the distinct impression that Camille was strictly a white folks' hurricane.

Mississippi Coast, 1959–68

On a bright spring day in 1959, the Gulfport police received a telephone complaint that a group of black children were using the beach. When a pair of officers went to investigate, they found a white nun—a teacher from the local black Catholic school—supervising a field experience for her science students. There wasn't another soul on the beach for a quarter mile in either direction. The police summarily sent the teacher and the students back to their school with a warning. Blacks weren't allowed on the Mississippi beaches.

The incident infuriated Dr. Gilbert Mason, a young black physician who had recently started a private practice in Biloxi. Mason was an avid swimmer, having had the good fortune to grow up near a decent swimming hole in Jackson, and he was strongly committed to providing wholesome physical activities for black youth. On Thursday, May 14, 1959, he and eight other blacks defiantly went for a swim at Biloxi Beach. The county's trash receptacles were in full view, confirming that the seashore indeed was public property. Police arrived and escorted Mason to headquarters, where he was given a warning but was not booked for any offense.

Dr. Mason's follow-up attempts to get an explanation from Mayor Laz Quave proved unsuccessful. Quave claimed only that the statute prohibiting blacks from using the beach was "on the books" but failed to produce any such documentation.

Mason did some research on his own and discovered that, although the natural tidal flats on the Gulf side of the road had originally belonged to the roughly two thousand corresponding properties north of the highway, the beach had become public when the county accepted federal funds for its improvement following the hurricane of 1947. And because it was public, of course the beach had to be legally accessible to blacks as well as whites. To file a legal petition on the matter, however, somebody would first need to be arrested and formally charged with the offense of trespassing. Mason commiserated with the local black leaders, gained their support, and called for a "wade-in." The event was scheduled for 1:00 p.m. on April 17, 1960—Easter Sunday.

When the appointed time came, Dr. Mason arrived at the beach alone. He waded into the water alone. He alone was arrested. Nobody else, of the hundreds of blacks who had pledged their support, showed up.

At Mason's Monday evening arraignment, however, there was a significant black presence. The word quickly spread through the black community that a leader had emerged in their midst, and another wade-in was scheduled for April 24. This time, 125 blacks participated.

Unfortunately, several hundred angry whites also showed up, and when the police cars drove away en masse, all hell broke loose. The violence wasn't confined to the beach; over the next few nights, there were drive-by shootings and acts of vandalism and arson in the black neighborhoods. One of two blacks killed was a young mentally retarded man found nearly decapitated, his body dumped on the median strip in front of the Jefferson Davis mansion—a symbolic message that was not hard to translate. Although the vast majority of the injuries were sustained by blacks, it was only blacks who were arrested. Nobody was ever detained for the two murders, the arson, the beatings, or the vandalism of black-owned property.

Mason's initial attempts to resolve the beach access issue in the state courts were stymied by stall tactics. Medgar Evers and the NAACP legal defense team got involved and, in a clever end run, petitioned the new civil rights division of President Eisenhower's Justice Department. On May 17, 1960, the U.S. Justice Department filed suit in federal district court against Harrison County, the board of supervisors, Sheriff Curtis Dedeaux, the city of Biloxi, Mayor Laz

Quave, and Biloxi police chief Herbert McDonnell. Their infraction: denying Negroes the use of the beach.

The battle, however, had barely begun. A group of southern segregationist political leaders met in Biloxi to rally the local whites to fight the federal lawsuit, and the keynote speaker was none other than "Judge" Leander Perez, the political boss of Plaquemines Parish, Louisiana. Perez pointed out how the parish *he* ran in the neighboring state knew how to keep blacks in their place. He urged all white Mississippians to rally behind their governor, Ross Barnett, warning them that if they didn't, "you may as well close your shops and hotels, barricade your homes, and keep your wives and daughters off the front porches." Thus energized, the defendants in the beach discrimination lawsuit managed to drag out the court proceedings for another eight years, and it was only on August 16, 1968—just one year before Camille—that the federal courts finally affirmed, almost anticlimactically, the right of every person regardless of race to have equal access to the twenty-six-mile stretch of beach in Harrison County, Mississippi.

That right, however, was paid for dearly over the previous decade. Dr. Mason received hundreds of threats, had his office firebombed and two of his cars torched, and was the target of an assassination attempt. Many of the black plaintiffs in the lawsuits were summarily fired from their jobs. Some blacks who hadn't even been involved in the wade-ins lost their jobs because of their personal associations.

The black community struck back by boycotting the hardware store that had sold the chains the white mob used to beat up blacks in the April 24 riots and by shunning all Borden products after that company fired several black workers for their alleged sympathies with the wade-in. The boycotted hardware store was forced out of business, and when unsold Borden products began perishing on store shelves, the company was forced to pull out of Biloxi.

The state-sanctioned culture of racism and intimidation of blacks ran so deeply in Mississippi in 1969 that few blacks would have felt comfortable telling reporters about their hurricane experiences. Even today, the older blacks still have little to say about the disaster; most simply shrug and say the hurricane was something that came and went. In life's grand scheme of things, and against the retrospective background of the segregation and racism of those times, Camille apparently was only one of a string of adverse experiences.

CHAPTER 11

RUBBLE

Pass Christian, Monday, August 18, 1969

Jackson M. Balch, manager of the Mississippi Test Facility (now the John C. Stennis Space Center), hugged his two young sons and led them down the hall. Although he'd primed himself mentally to find damage downstairs, the actual sight still astounded him. Most of the staircase was ripped away, and the walls of the first floor were essentially gone. The second floor was perched precariously on a bare skeleton of open studs that threatened to collapse at any moment. This, despite the fact that the home stood twenty feet above sea level.

Although he ran a big show at the space center, with responsibilities for 2,682 employees and three government contractors, Jackson had a hard time getting his thoughts organized. The few local officials he met were likewise in a daze, confused about where to start, what to do, how to do it. No communications whatsoever were reaching Pass Christian, and no physical resources were available—not even something as basic as drinking water. In fact, most of the infrastructure had been destroyed throughout all of southern Mississippi, and even the carefully planned emergency response mechanisms were in ragged shape. Help would be slow in arriving.

With the sun promising a sultry day ahead, Jackson decided that he was probably in as good a position as anyone to try to get things rolling. He took his sons and walked to the bay. Debris was piled twelve feet deep for a stretch of 150 yards at the east end of the bridge; clearly no help would arrive from the west. Beyond the pile of jetsam, the pavement slabs were displaced crazily on the eastbound lane. They climbed over the rubble and made their way across the bridge to the town of Bay St. Louis, which was also in a state of serious distress. Borrowing a car that had been parked high enough to avoid the floodwaters, Jackson drove the remaining fifteen miles to the space center, zigzagging around one obstacle after another.

The Mississippi Test Facility had taken in twelve hundred refugees prior to the storm, and many of those locals were still there. Although none of the buildings had suffered structural damage, few windows remained unbroken anywhere, and that included the windows in the vehicles in the parking lots. The culprit was the gravel that had been used to weigh down the flat membrane roofs; Camille's vicious winds had scooped it up and blasted it into every object in the vicinity. When the eye passed through and the winds reversed, some of this same gravel was scooped up again and slung in the opposite direction.

The secure underground phone line remained alive, and Jackson called his boss at the Marshall Space Flight Center in northern Alabama. That "boss" was Dr. Werner von Braun, the former Nazi rocket scientist who had headed the development of Hitler's V-2 rocket program at Peenemunde, Germany, and who had thereby contributed to the deaths of hundreds of London civilians during World War II. The man who had introduced the world to a whole new category of weapons of mass destruction, however, was also known to have a human side. When the first V-2 hit Chiswick, a London suburb, on September 8, 1944, he had remarked to his colleagues, "The rocket worked perfectly except for landing on the wrong planet." Before the Allied capture of Peenemunde, von Braun negotiated the surrender of five hundred of his top scientists and engineers, along with their missile blueprints and several actual test vehicles, to the Americans. In 1960, von Braun was appointed director of the Marshall Space Flight Center in Huntsville, where he supervised the development of the Saturn V launch vehicle that had propelled three Americans to the moon in July, just one month before Camille.

Werner von Braun's first questions to Jackson Balch were not about the condition of the Mississippi Test Facility but about the nearby communities. Jackson replied, "You were in Germany under the Allied bombing raids, and I'm sure you well remember the devastation. That's what Pass Christian looks like." As for emergency assistance, Jackson explained that there wasn't any, at least not yet, and that nobody seemed to be in charge. Von Braun didn't take long to mull over the matter. He told Jackson to do whatever he thought appropriate and that Huntsville would ship some provisions directly to the community. The former Nazi was true to his word; the next day, thirty NASA trucks rolled into Pass Christian with chainsaws, generators, medical supplies, and other necessities, courtesy of a man who had started his own career destroying cities and who was stepping far beyond the sphere of his administrative authority in sending relief to the stricken Gulf Coast.

After stuffing as many ten-gallon containers of water as would fit into two NASA security patrol cars, Jackson and another NASA employee headed back to the bay bridge. The sultry air hung thick and humid. Setting up a human chain of volunteers, he painstakingly transferred the water across the wrecked span. By then it was getting dark. Not far from the buried east end of the bridge, he spotted the headlights of two jeeps that had gotten there by driving along the beach. The vehicles were manned by four National Guard enlistees with orders to halt all bridge traffic in the interest of preventing looting.

Jackson was incredulous. It was still another mile to the center of town, they were lugging eighty-pound cans of water, and here were two jeeps and four men being squandered to prevent *looting*? He realized, of course, that it wasn't the fault of the young troops themselves; they were simply obeying a stupid order. Jackson, himself a former member of the Guard, assured them that there was no way in hell that any vehicle was going to cross that bridge, nor would any self-respecting looter take the trouble to make such a trek on foot just to root around in the worthless rubble at this end. Meanwhile, the townsfolk desperately needed water and other supplies. The young soldiers relented and loaded the water containers into their jeeps.

Some of the locals, working mainly with their bare hands, had already cleared a path into town that was wide enough for a vehicle to squeeze through. Jackson and his group drove to St. Paul's Church school, which was being used as a field hospital. He approached a nun and asked her what they were doing for the victims.

"We are praying," she answered.

"Dammit, sister," Jackson retorted, "this is no damn time to pray. Let's get to work!"

Five months later, Jackson Balch would testify before a U.S. Senate subcommittee investigating the efficacy of the disaster relief following Camille. After being recognized by the chair, Senator Birch Bayh of Indiana, Jackson stated for the record:

> Mr. Chairman, I can do one of two things. I can read a sanitized statement which has been approved for me to read, or I can give you a picture from being a participant of [my own] experience about what the story is.

Bayh responded that the sanitized version would be put into the record as if it were actually read. Then he said, "I, for one, would like you to push your chair back and tell us, as seen through your eyes, what would be helpful."

Jackson Balch was not the kind of man to pass up an opportunity like that. The next thirteen pages of the published fine-print transcript are packed with Jackson's comments—a total of about seventy-six hundred words—in which he addresses the lack of communication and coordination; the lack of preparation for the simultaneous failure of the water, sewerage, and transportation systems; and, most scathingly, the decision-making apparatus in the command structure of the National Guard. He ridiculed the concept that Guard units should be deployed to prevent looting when the victims themselves had no water, no medical attention, no transportation, and no sanitation facilities. He pointed out that key decisions needed to be made by the local authorities—even if they were ad hoc authorities—who had the most familiarity with the immediate human needs, and not by some officer many miles away who was ignorant of the actual situation. To the embarrassment of some of those present, he told the story of how Dr. Werner von Braun had been more responsive than many regional and state organizations, and he admitted with full candor that he suspected that his comments were likely to get him into trouble.

He described how he went to Gulfport that Wednesday to meet the regional commanding officer of the Guard and succeeded in getting a platoon assigned to search and rescue duties in Pass Christian under his personal supervision. To help those white boys find their way around town, he recruited fifteen black volunteers to ride with them in their jeeps, a unilateral decision on his part that raised some eyebrows. Meanwhile, in a similar no-nonsense fashion, Jackson brought in some of his rocket engineers to tramp around town shutting off gas leaks and evaluating what it would take to provide basic sanitation facilities. As for making urgently needed repairs to the schools—those that had survived—he simply went ahead and hired the labor using his NASA budget, figuring that the accounting mess (and indeed it was a mess) could be sorted out later.

Although Jackson Balch's main point in his testimony was that disaster response needs to be coordinated by the locals in a stricken region, some of the senators seemed to hear the exact opposite message. What was needed, in their view, was a more centralized national program for disaster management.

If people were to draw their conclusions solely from Monday morning's *Times-Picayune,* they would deduce that Camille had been a minor storm at best. There were photos of a light plane flipped over at the New Orleans airport, images of storm damage to awnings in the business district, and sketchy reports that a few city streets had flooded. There were human interest stories from the city hurricane shelters and, for the first time, a comprehensive list of those shelters. On an inside page appeared a photograph of a Weather Service meteorologist pointing to a storm map and predicting Camille's landfall near Fort Walton Beach, Florida. As for southeastern Louisiana, the only news copy was a brief mention that there was no news. However, 908,100 acres of Louisiana lay flooded, and so much of the lower peninsula had been totally obliterated that no complete census would ever be made of all the buildings destroyed there.

Before Monday's paltry edition was even off the presses, reporters were swarming to the Mississippi coast to remedy the deficit of information. The most obvious physical feature was the mammoth debris line roughly four blocks inland in Pass Christian, decreasing to three, two, then one block in the towns and cities to the east. Sections of Highway 90 were gone, and barges and other wrecked marine equipment lay scattered all along its twenty-six-mile beachfront right-of-way. Historical architectural treasures did not fare well; hundreds of them were washed away or flattened by the storm. Beaulieu, a mansion built in 1854 and referred to as the "Dixie White House" after a seventeen-day stay by Woodrow Wilson in 1913, was reduced to rubble. Only the foundation remained of Pirate House, a large colon-naded home dating from 1802 that reputedly served as a secret refuge for the pirate Jean Lafitte. Jefferson Davis's Beauvior had been damaged, and its grounds and outbuildings were in shambles.

As those sad images and reports began to appear in the media, readers and viewers got the impression that the destruction had been confined to the beachfront strip along the Mississippi coast. The devastation was much broader, however, and it extended into places most folks knew nothing about or, if they did, seldom thought of. When the Mobile District of the Army Corps of Engineers conducted a study of the inundated areas within its jurisdiction, it discovered that the massive flood had circled behind the coastal towns and cities to sow havoc as far as twenty miles inland. A twenty-eight-foot tidal surge does not merely stop when it piles rubble up to its own height; it seeks alternative ways to relieve itself, and the beachfront was not the only low-lying land in the region. High water had surged up rivers, tributaries, and bayous, destroying homes, forests, pastures, farms, and livestock. And not just in Mississippi, where 211,900 acres of land were flooded, but also in Alabama, where flooding inundated another 209,100 acres.

There were no historical architectural treasures in these obscure inland regions. Nor did those places have any prominent families with the political clout to garner attention. The victims of inland flooding were mostly poor blacks and poor whites, not necessarily living side by side but nevertheless sharing the same physical geography. Their plight was uniformly invisible in the newscast flyovers, ignored by the politicians, and unaddressed by the relief organizations that began arriving on the coast.

In the first few days, the main problems were water, medical attention, sanitation, and shelter—in that order. Food would become an issue later, but initially it was plentiful; anyone with the means to open a can could find something to eat. In the high-ground neighborhoods where many homes survived at least partially intact, steak, lobster, shrimp, and frozen vegetables emerged from thawing freezers and refrigerators to be prepared on open fires fueled by the sudden abundance of firewood. In an ironic reversal of class status, quality food was more abundant in the middle- and lower-class urban neighborhoods than along the streets of the devastated upscale man-

Erosion on Scenic Drive in Pass Christian. The beach and the seawall, having been submerged so far below the breakers, suffered minimal damage. *(Photos by Bill Fennell.)*

sions nearer to the coast. The most expensive perishable food, of course, was eaten first. Unfortunately, some of the post-disaster relief workers would misinterpret this logical practice as evidence of people depleting their food caches in anticipation of expected handouts.

Twenty-six counties (or parts thereof) in Mississippi, two counties in Alabama, and nine parishes in Louisiana were declared disaster areas, but the grim statistics of damages and deaths would not be tallied until months later. The Army Corps of Engineers would ultimately report that in the Mobile District (Mississippi and Alabama) flooding had destroyed 3,566 homes and damaged 16,250 others, while wind and fallen trees had been primarily responsible for destroying another 302 homes and damaging 25,842. In addition, a total of 325 Mississippi businesses had been destroyed and 1,825 had been damaged. The surveyors noted that the numbers should be considered approximate, insofar as it was impossible in many cases to attribute the storm damage specifically to flooding or to wind. Such a detailed census of damaged structures would not even be attempted in Louisiana.

Tides surged to record high levels from Bay St. Louis (21.7 feet) to Pascagoula (11.8 feet) and along the Alabama coast from Bayou La Batre (8.5 feet) to Dauphin Island (9.2 feet). Even Mobile had a tide of 7.4 feet, while a three-foot surge was reported in Apalachicola, Florida—350 miles east of the landfall. The populations of the flooded areas alone numbered 27,000 in Louisiana, 53,300 in Mississippi, and 7,500 in Alabama; these figures stood in addition to uncountable tens of thousands who suffered wind and rain damage only, which in many cases would never be claimed or even reported to the data gatherers. The death toll was reported as 9 in Louisiana and at least 172 in Mississippi, with the caveat that dozens of others remained unaccounted for and that it was not always possible to attribute a particular death to the effects of the hurricane. Overall, however, the low ratio of fatalities to property damage was consistent with other natural disasters of twentieth-century United States. Many homes that would be declared total losses had in fact successfully sheltered their occupants from serious injury during the fury of the storm.

To the amazement of the civil engineers, Camille left the Mississippi coast's long artificial beach relatively intact, and even the concrete seawall suffered only small patches of damage. Apparently the surge had arrived so quickly and had inundated the shoreline features so completely that they were never subjected to the scouring action of the waves. Although fallen trees brought down power lines and obstructed streets throughout the region, at least 90 percent of the trees survived, denuded of leaves and with branches missing but still rooted. This survival rate for trees stood in stark contrast to the man-made structures, most of which were destroyed or seriously damaged. Clearly, Mother Nature had known what she was doing when she designed trees to flex in the wind. Of the 10 percent of the forests that were flattened, including nearly forty thousand acres of tung trees and 1.2 billion board feet of sawtimber, about 85 percent was ultimately salvageable.

Offshore, nine barrier islands were completely washed over and a tenth one was 70 percent inundated. Eight of those flooded islands permanently lost a total of 542 acres, one actually gained 8 acres, and one—Pelican Island—completely disappeared forever.

The three freighters grounded at Gulfport notwithstanding, by far the worst damages to shipping were in Louisiana. Although the Army Corps of Engineers reported that at least ninety-four vessels sank or ran aground in the Mississippi River during Camille, the detailed listing is incomplete. There were nine seagoing ships grounded near Pilot Town. An 8,985-ton Norwegian tanker, *Stolt Orator,* was stranded in Northwest Pass, which, if accurate, is a curiously inappropriate little channel for a large ship to end up in. The 13,080-ton Brazilian tanker *Mario D'Almeida* went aground in the Southwest Pass, where it shouldn't have been without a pilot aboard. Other than this, there is scant information about most of those sinkings and groundings in the river.

As for the dozens of ships that were clustered out in the Gulf beyond the jetties, there is even less information. The tug *Charleston* sank east of Southwest Pass; the barge it had been towing, the *City of Pensacola,* washed ashore and was recovered. The 9,294-ton *Yamawaka Maru* ran aground near the mouth of the river after tangling with an offshore oil platform, heavily damaging both. But the Gulf of Mexico is outside the jurisdiction of the Army Corps of Engineers, and even today there exists no master depository of information relating to hurricane damage to Gulf shipping. One thing is certain: August 17, 1969, could not have been a pleasant night for the crews of any of the ships in the waters around southeastern Louisiana.

Offshore, as of daybreak Monday, one oil well was out of control and two drilling rigs rested on the sea bottom. Several platforms had been scoured of every piece of surface equipment. Four rig tenders were damaged to the extent that they were inoperative, and one barely remained afloat.

As engineers and oil workers sprang into action to contain the physical damages and the small spills, the oil company executives sought to control the public image damage. Tight-lipped executives found dozens of reasonable-sounding excuses not to talk to journalists, and those reporters who were out to get sensational stories had no lack of other people to talk to. Thus it was that only the skimpiest information about the fate of the roughly six hundred affected oil rigs and platforms ever reached the press or the broadcast media.

Fresh in the oil executives' minds was an accident that had

occurred less than seven months earlier—on January 29, 1969—not in Louisiana but on a platform six miles off the coast of Santa Barbara, California. There, the Union Oil Company was drilling a deep well thirty-five hundred feet below the ocean floor using casings that failed to meet federal or state standards. When riggers began to retrieve the drill pipe to replace the bit, they had a natural gas blowout. In the eleven days it took to cap the rupture, which created five separate fractures in the ocean floor, two hundred thousand barrels of crude oil bubbled to the surface. Winds and swells spread that ooze into a slick covering eight hundred square miles. As the more volatile components evaporated, thick, sticky tar washed up along thirty-five miles of pristine California coastline. When the press interviewed Fred L. Hartley, president of Union Oil, he replied that he didn't consider it a disaster, because no human lives were lost. Then he unwisely added, in a widely publicized remark, "I am amazed at the publicity for the loss of a few birds."

If there was any event that would mobilize the nascent environmental movement of 1969, that was it. Scruffy long-haired hippies scoured the beaches to rescue the mired feathered creatures, and millionaires used their luxury cars to transport tar-covered grebes to the emergency treatment centers at the Santa Barbara Zoo. Donations to environmental organizations soared as dozens of lawsuits were filed in California courts. Although it would be several years before all of those litigations would be settled, one lesson became clear to every oil executive in the country: the industry couldn't afford too many other incidents like the one in Santa Barbara, and it certainly couldn't afford to act flippant about even the smallest spill.

The Santa Barbara ecological disaster had been a singular incident: one platform, one accident, one big spill. Now, in the wake of Camille, at least six hundred offshore oil platforms had been struck by winds and waves of an intensity that nobody had ever thought possible. Although those platforms had been designed to withstand a twenty-five-year storm (their decks were thirty-five to forty feet above the water, and their superstructures were designed to withstand wind speeds of around 135 miles per hour), Camille's brutal punch had been considerably in excess of those design criteria. Given that the storm surge was more than twenty feet, it was likely that many waves had towered more than forty feet above sea level—higher than the lowest decks of most of the platforms.

If a drilling accident in perfect weather had led to a spill of two hundred thousand barrels in California, what environmental disaster was Louisiana in for within the next few days? The state of Louisiana produced an average of about two hundred thousand barrels of oil per *day,* much of which came from the Gulf and a good fraction of which came from the rigs that had been struck dead-on by Hurricane Camille.

For the oil companies, the immediate order of business was damage assessment, first by helicopters carrying company engineers. Unfortunately, all of the heliports in Plaquemines Parish were still underwater and were likely to remain that way for quite some time. Flights would need to originate from farther inland, and they'd need to be of short duration, respectful of the fuel needed for the extended trips. Company managers in Louisiana and their executives elsewhere in the country nervously waited for the news, hanging on every phone call from New Orleans as the reports dribbled in slowly over the next few days.

From the air, it was easy to see that several platforms were simply gone and dozens of others were seriously distressed. Remarkably, however, there was only one oil slick visible from the air. Repair equipment was immediately dispatched to that site, and the spill, apparently about five hundred barrels, was quickly contained. Thousands of safeguards and secondary and tertiary backup systems had apparently worked perfectly under the most grueling possible conditions.

It would be many weeks before the offshore damage census was complete. When it was, the results were remarkable. Fifteen drilling rigs or full-production platforms had been totally destroyed, and another ten had been seriously damaged but were repairable, out of several hundred that had been in the hurricane zone. Several floating rigs had borne the worst of the storm with little or no damage at all, including, remarkably, the Odeco rig that had been towed from Mobile to avoid the hurricane and instead was struck almost dead-on. All of the personnel had been evacuated from every stationary platform in a timely manner, and there had been no loss of life. The oil companies had also contributed to scientific knowledge about the hurricane, for it was that recording anemometer on a Transworld rig thirty miles east of the storm track that confirmed that sustained winds had reached at least 172 miles per hour. Yet it would not be a matter of quickly getting back to business as usual. It would take several months to properly inspect the structural integrity of each of the surviving rigs and platforms, in many cases using divers, before resuming full operations.

The oil companies never did go public with much of the news of how well their offshore installations had fared. They revealed the minimal skimpiest data sought by the Army Corps of Engineers for its study, and they never divulged much of anything else that had happened. One hopes that they took this tack out of an altruistic belief that a natural disaster of Camille's scope should not become any company's public relations event. On the other hand, there are many people, even today, who would be reassured by learning that, even as early as 1969, the oil companies were not as uniformly irresponsible as they were portrayed following the 1969 Santa Barbara spill or, more recently, after the Exxon Valdez ecological catastrophe of 1989.

Keeping such information tightly under wraps is no longer the practice. The Interior Department's Mineral Management Service (MMS), as well as the private National Ocean Industries Association, publicized a large amount of information after Hurricane Andrew walloped hundreds of oil platforms in 1993. In that event, fewer than 8 percent of the offshore rigs had been damaged, and only a handful of the older ones were destroyed or damaged beyond repair. As in Camille, the total oil spilled amounted to just a few hundred barrels, all of which were quickly contained.

Today, there are 3,439 oil platforms operating off the Louisiana coast, and new ones are being added in deeper waters as old ones are decommissioned and dismantled. It is a large industry prone to numerous risks, but one that has in many respects, at least regarding hurricane hazards, risen to the technical challenges and standards of environmental responsibility thrust upon it by an oil-hungry nation.

From the published accounts, it's easy to get the erroneous impression that swarms of looters invaded Plaquemines Parish, the Mississippi Gulf Coast, and even Nelson County, Virginia, in the disaster's aftermath. Indeed, there is scarcely a longtime resident of any of these three regions who does not have stories to tell about such pillaging. Video footage shows haggard survivors holding shotguns while guarding the wreckages of their homes. Other ruined structures are spray-painted with signs promising that looters will be shot on sight.

One victim surprised several men picking silverware out of the wreckage of his home, and they quickly dispersed when he threatened to shoot them. A merchant was caught selling canned goods he'd recovered from the debris; in that case, the sheriff had the goods confiscated by the health department. Another man learned that one of his rifles lost in the flood was in a neighbor's possession; he confronted the neighbor, and the gun was sheepishly returned. The main items that people claimed to have been looted were jewelry, firearms, and tools.

Direct experiences, however, were the exceptions. Most of the looting stories amounted to someone hearing that someone said that someone saw something, and so on. In fact, official crime rates invariably *drop* after a disaster. Nor does there seem to be a single incident of a looter being shot, not just in the wake of Camille but after any U.S. natural disaster in the last fifty years or so. This, in a nation where peacetime assaults with firearms are the highest in the world.

There are reasons for the widespread concerns about looting, some obvious, some a bit more subtle. In terms of human psychology, not only are natural disasters violations of our sense of control over our own living space, but, more fundamentally, they violate our confidence in the resiliency of our society. Humans have succeeded as a species not because of great physical strength or fleetness or eyesight or any other physical characteristics; our species has prospered because we have developed the ability to anticipate and plan for the future. And how does a human society plan? Only by paying attention to the patterns of the past. If traffic jams grow more frequent, we plan to widen the roads. If our schools get overcrowded, we plan to build new ones. If our rivers flood repeatedly, we plan to contain them or to divert the excess flow. And so on.

Because disasters are discontinuous events that follow no discernable pattern, they undermine their victims' confidence about the future. Those hit the hardest often feel especially vulnerable to further upsets in their personal lives, and this compels them to exercise the maximum control they can over what happens next. Tempers do flare, and signs indeed appear saying things like "Looters Will Be Shot." But do disasters actually transform hordes of normal folks into vigilantes and thieves? The evidence suggests that they do not.

There is also a fair amount of ambiguity regarding what constitutes looting. Few people would consider picking a muddy twenty-dollar bill out of a garbage pile to be looting. The law explicitly recognizes that the salvage of a sunken ship is not looting. If someone notices a shiny object in the rubble being transferred to a burn pile by a high-end loader and stops to remove some pieces of jewelry, is that looting? What about taking a crowbar or a chainsaw from someone's garage because it's needed to free trapped victims?

Undoubtedly there will always be some degree of pillaging after a major disaster. Seldom, however, does it extend to items of any great value, and in most instances the "looting" label is a matter of interpretation. As for violence in the wake of a disaster, the general pattern is that the worse the destruction, the fewer the physical assaults in the following days. If there is any event in society that brings out the best in human behavior, at least in the short term, it is a major catastrophe.

Mississippi Coast, Tuesday, August 19, 1969

Hubert Duckworth was far from optimistic. Even the national news, with Walter Cronkite no less, had reported that no one had survived the destruction of the Richelieu. Now that he'd seen the site of the apartments firsthand, he knew in his gut that it would take one hell of a miracle to outlive devastation like that. He dismissed Mike Gannon's reassuring words that his son had survived as a misguided act of kindness to an apprehensive father. Realistically, he could hope only for enough luck to find Ben's body.

His son-in-law, Bill, and Ben's buddy Charles Edward pleaded with him to keep his faith. The network news might be wrong. It often is. Ben's a strong fellow, an excellent swimmer. They're gonna find him alive.

Hubert nodded without responding. Inside, his soul was churning.

They tramped to the high school. Muscling through the crowds at the message exchange boards that had spontaneously sprung up, Charles checked the postings. Nothing from Ben. He scribbled a

Laundry in the trees. Such sights were common and led to numerous erroneous reports of human bodies dangling in the branches. *(U.S. Navy.)*

note of his own—"Ben Duckworth Call Home"—but then he crossed out "Call Home" and replaced it with "Post Message—We're Lookin for You!"

Various relief activities were in progress, from first aid administration to the dispensing of potable water. Record keeping was nonexistent. Hubert asked around about Ben. Nobody of that description was there, although if he was injured, he may have been taken to a hospital. Unless, that is, he was . . . uh . . .

The temporary morgue was in the shop area. More than a dozen sheet-covered corpses lay in a row on the damp concrete floor next to several ruined antique cars. Only one of the bodies was of the right gender, age, race, and size to be a prospect. The attendant peeled back the shroud. It wasn't Ben.

It was the kind of failure that generates a glimmer of hope. For the first time, Hubert began to entertain the possibility that Mike Gannon's reassurances were authentic. Maybe, just maybe, Ben had indeed survived.

It was evening when Ben opened his bleary eyes from another of his near-continuous series of naps on the sofa in the navy medevac office and saw a sight he'd never seen before. It was his dad crying, and his dad never cried. When Ben sat up, his dad hugged him, and his dad never hugged. Behind Hubert stood Bill and Charles Edward, the latter clapping and shouting, "I knew it! I knew it!"

"Come on, son," Hubert said, patting Ben's back as he wiped his tears on his other sleeve. "We're goin' home."

CHAPTER 12

DELUGE

A hurricane ceases to be a hurricane when its sustained winds drop below seventy-four miles per hour, at which time it is downgraded to a tropical storm. When the winds decline further to under thirty-nine miles per hour, the storm is reclassified as a tropical depression and is stripped of its name. Shortly after entering Tennessee around noon on Monday, August 18, as dazed Gulf Coast residents searched for loved ones, the remnants of Camille became just one of many other unnamed weather events that sweep across the nation from day to day in the general direction of west to east. Now the responsibility for tracking the system's progress rested solely with the National Weather Service, and there would be no further input from the National Hurricane Center.

As the clouds drifted over Kentucky, parts of that state received a smattering of rain: one-half to one inch in most places, with no flooding. On Tuesday evening, August 19, the Weather Service reported that the residual storm was expected to wane and "disappear over the Appalachians early Wednesday." At 5:20 a.m. on August 20, an early arriving weatherman at the Louisville station reported that the rain indeed had ended and "the remains of Camille did a beautiful job of watering the farms and lawns across Kentucky yesterday." Few meteorologists in the country had any hint of the catastrophe that had struck Virginia a few hours earlier.

Despite the showers in Mississippi, Tennessee, and Kentucky, most of the 108 billion tons of moisture Camille had vacuumed from the Gulf remained in the sky. The Blue Ridge Mountains deflected those clouds upward, which alone was enough to increase the rate of condensation and to trigger heavier than usual precipitation east of the mountains. But by a cruel coincidence, something else was also happening: a cold front—a giant ground-hugging wedge of cold heavy air—was moving in from the north. The two meteorological systems were on a collision course, and they would converge (more or less) over Nelson County, Virginia, where the summer air already hung thick with humidity.

For the next several decades, meteorologists would speculate and disagree about the precise mechanism that precipitated the great Nelson County disaster. The tail ends of hurricanes had traveled this same general route many times before, and although those storms sometimes dumped a significant amount of rain east of the mountains, nobody had ever witnessed the likes of the deluge on the evening of August 19, 1969. The meteorological models of the time were too coarse to analyze microweather events that arose in a region only a few dozen miles across. Yet if all of the hindsight investigations by the weather scientists were inconclusive about why the Virginia disaster took place, they did at least relieve the meteorological community of its culpability for not having issued an advance warning. Given that the experts were still arguing decades later about the causes of the catastrophe, they scarcely could have been expected to *predict* it, or at least so the reasoning went. What happened that terrible night came as a total surprise—not just to the victims but also to the weathermen.

Although rain clouds in these parts normally sweep over the mountains in a relatively smooth sheet, on that particular summer evening of August 19, 1969, observant locals noticed something strange about the sky. Instead of simply spilling over the ridge, the oncoming clouds were tumbling and rolling, growing darker by the minute, and billowing ever upward into a huge cauliflower-shaped monster. Within that rapidly growing thunderhead, the turbulence was obvious and severe. Yet the cloud mass did not move on; instead it stalled over Nelson County and continued to grow and boil like a witch's brew. Some said that the sky had a weird grayish yellow hue, while others said it was almost black. To everyone who gazed skyward, regardless of their vantage point, it was clear that these were no run-of-the-mill rain clouds. At ground level, however, there was dead calm—no wind at all. Folks indoors had no inkling of the fury developing over their heads. All they knew was that it was beginning to rain. Again.

Nelson County, Virginia. *(Adapted from Virginia Department of Transportation.)*

It had been an uncharacteristically wet August in Virginia. One frontal system after another had withheld its moisture from the parched fields of Kentucky and released its rainfall on the eastern slopes of the mountains. The hillside soil, typically no more than two feet thick overlying the 350-million-year-old sandstone, was already soggy. Yet none of those prior rains had been disastrous. Propelled by the steep drop in elevation from the Blue Ridge Mountains to the James River, the numerous creeks and streams ran swiftly, and they efficiently channeled excess rainwater out of the county. Except for the banks of the James River itself, flooding in Nelson County was rare, and those few times when a stream had overflowed within living memory, it had been more of a messy nuisance than a threat to lives and property.

Mother Nature, however, operates on geologic rather than human time scales, and this region had indeed experienced disastrous flooding before, at intervals of every ten thousand years or so. The thick rich soil around the village of Massies Mill was proof of major flooding in ages past. Up in the mountains, more dramatic evidence was about to be revealed that such prehistoric floods, albeit infrequent, were sometimes unimaginably violent.

Massies Mill, Virginia, August 19–20, 1969

Fourteen-year-old Warren Raines heard the phone ringing in the wee hours of the morning. He and his siblings scrambled to his parents' bedroom to investigate what news could possibly be important enough to arrive in the middle of the night. The caller was a woman from a few blocks up the street, warning the Raineses that the forty-home village was flooding. Warren's father assured her that she was overreacting; the nearby Tye River, after all, was just a gurgling creek that seldom got more than a foot or two deep and that would have had to swell to many times its normal size to top its banks. The caller, however, insisted that her car had already floated away. Warren's mother peered out a second-floor window and then beckoned to everyone to take a look. Several cars were crawling south on the main street, Route 56, already covered with water.

The phone rang again. The Woods family, neighbors toward the river, asked whether the Raineses were leaving and, if so, whether they'd take the four Woods kids with them. Mrs. Woods was confined to a wheelchair, and it wouldn't be wise to take her out in a flood, but they wanted to assure that their children would be safe.

Carl Sr. agreed. Then he noticed that water was approaching their front door. In what would turn out to be a tragic misjudgment of the urgency of the circumstances, the Raineses consumed precious time moving furniture upstairs before Carl Jr. pulled the family station wagon around to the front of the house.

Eleven people—two adults, five teens, and four youngsters—stuffed themselves into the vehicle. They'd traveled less than a block when the engine quit and refused to restart. The water was more than a foot deep and growing swifter. Carl Sr. ordered everyone out, figuring that they'd wade down the road a quarter mile to a friend's home on higher ground. The rain fell in buckets, and Carl Jr. passed his little sister Ginger to their mother in a seemingly inconsequential act.

At the hamlet of Tyro, a few miles upstream, the situation was already grim. There, hundreds of trees were losing their footing on the saturated slopes and sliding into the narrow mountainside streambed. Tangled masses of timber dammed the river at several spots but only for a few minutes at a time before the pressure of the rising water burst those logjams. Houses and outbuildings were no match for the angry explosions of debris-laden floodwater, and the avalanches swept all human constructions from their foundations and ground most of them to smithereens.

Even as the Raines entourage waded along the flooded main street of Massies Mill, at least nine of their friends and neighbors upstream in Tyro were already dying in the cascade. One of those upstream victims, seventeen-year-old Audrey Zirkle, had been chosen Miss Nelson Teen just a year earlier.

With safety within the Raineses' view, the floodwaters surged. They tried desperately to hold onto each other by locking arms, but it was impossible. The fierce current separated them and swept them downstream.

Warren snagged a cluster of vines. In the nearly continuous lightning, he saw his mother clutching a tree limb a hundred feet downstream. The vines began to disintegrate. She yelled to him to let go; she'd grab him when he floated past. He remanded himself to the angry water for a few breathless seconds. When he surfaced, his mother was gone. The last words he heard from her were, "I'll catch you."

He grabbed a limb and worked his way to a more substantial-looking willow whose trunk immediately cracked. He clutched the splintered fragment of that tree, his head barely above the rushing water. All manner of debris raced by. Trees, light poles, cars, cows, parts of buildings. He ducked. He dodged. Hour after terrifying hour. An entire house sped toward him. He was about to dive and take his chances in the current again when at the last moment the structure spun away. The driving rain continued. Through the rest of that unforgettable night, young Warren Raines hugged what was left of the abused willow tree, dodging an unrelenting salvo of flood-borne debris.

Sheriff Bill Whitehead, age forty-three, was a steel-jawed man with a graying crewcut whose booming voice and six-foot-seven frame were enough to intimidate any outsider. Most of Nelson County's natives knew him as a selfless public servant with an unwavering dedication to the safety and well-being of the locals. Although his great-grand-father had been a slave trader who fought for the Confederacy during the "War of Northern Aggression," Whitehead was the first sheriff south of the Mason-Dixon Line to hire a black woman as a deputy. He created an outside work program for county prisoners, and when some of the state prison board criticized that agenda, he brusquely retorted that he didn't see anything wrong with letting a man earn a few dollars and some self-respect while serving his sentence. He talked the language of the local farmers, and his own farm, Willow-brook, with its handsome 1810 brick farmhouse, was one of the pret-tiest in the foothills.

Law enforcement in Nelson County was synonymous with the sheriff's office. In its entire 472 square miles, the county lacked a sin-gle local police department, and the Virginia State Police, although having jurisdiction there, maintained no physical facilities within the county's borders. If there was a crime or a civil emergency or even a dispute among neighbors, it fell to the sheriff's office to handle the matter.

Whitehead got home around nine o'clock that Tuesday evening, just about the time the rain began. A few hours later, one of his deputies phoned to report that the Piney River Fire Department, while responding to several calls to pump out basements, managed to get its pumping engine stranded away from the firehouse. A second caller informed him that high water and debris were blocking several roads and some homes were isolated. There was nothing that could be done right then, but the sheriff would need to get the situation in hand first thing in the morning. Whitehead set his alarm clock and turned in. Tomorrow might be a busy day.

Although a mere thunderstorm wasn't the kind of thing to keep the sheriff from sleeping when he had a mind to do so, his wife, Catherine, was another matter. Having experienced a tornado while growing up in northern Alabama, summer storms always made her restless. She got up, discovered that the power was out, rummaged around for some candles, and then noticed water pouring in over the bedroom windowsills. She shook her husband awake to help her mop it up.

All the windows in the other rooms were leaking just as profusely. As Bill went around opening the storm windows to let the water run out, he heard what sounded like a roaring wind. Yet the trees in his yard, illuminated by the near-continuous lightning, were limply still. He peered across the lower pasture toward Hat Creek, a tiny tributary of the Tye River. Cloud-to-cloud lightning lit the scene from horizon to horizon. To his astonishment, the brook had engulfed the field, and in its former channel white-capped waves thrashed to heights of six to eight feet. Although the terrain sloped away from the house, the rain couldn't run off as fast as it was falling, and the backyard had become a lake. As for the sound he had mistaken for a high wind, it would turn out to be a landslide a few miles upstream—one of the first of about 150 in the county that night.

He tried to phone his mother, who lived a short distance up the drive, but the connection was dead. By now, his two teenage sons and his daughter were awake, and although Whitehead correctly assessed that the flood wouldn't reach their home, he nonetheless directed everyone to keep their eyes on the water and to head to higher ground if that should become necessary. He then hopped into the patrol car with his eldest son, drove up the lane to check on his mother, and, after assuring himself that she was safe and paying attention, continued out onto Route 151 to investigate the broader region.

Two miles to the north, he encountered a washout and had to turn around. A quarter mile in the opposite direction, a little stream called Possum Trot had inundated the road. All of the unpaved side roads were also blocked by high water or debris. He pulled up to the Roseland rescue squad building and jumped out. Whitehead had to cup his hands over his nose and mouth in order to breathe as he ran through the horizontal rain. Inside, sitting out the storm, were two squad members and a state police sergeant. All three had gotten stranded on this two-mile stretch of road while trying to respond to calls for help.

Whitehead needed to get a handle on the situation—was this a case of a freak local flood, or was it part of more widespread trouble? The landline phones, however, were useless, and his sheriff's radio raised no response. He had long known about the shortcomings of the office's communication system in this hilly country. As would come out later during the February senatorial hearings, a "national radio quiet zone" blanketed much of the region, and this had precluded the development of effective radio transmission and repeater sites not only in Nelson County but also in neighboring areas.

The quiet zone had been established in 1958 to prevent terrestrial signal encroachment on the giant space-aimed radio observatories just over the state border in West Virginia. Ostensibly, those arrays of receiving dishes served the scientific needs of radio astronomers. They also, however, picked up signals from the early generations of weather and spy satellites and additionally served as a Cold War listening post for transmissions that might be intercepted from orbiting Soviet satellites. Officially under the jurisdiction of the Federal Communications Commission (FCC) and the Interdepartmental Radio Advisory Committee, the quiet zone was administered by the Naval Research Laboratory and the National Radio Astronomy Laboratory. One task of those organizations was to review all regional transmitter license applications, to conduct propagation tests to assure lack of radio interference, and to approve (or deny) any proposed changes in local radio transmitting facilities. In executing these duties, the administrators placed a cap on the effective radiated power of repeater transmitters in the region at a paltry one watt. This drastic restriction effectively precluded the construction or use of repeater stations in the region.

In 1965, Rockbridge County (contiguous to Nelson County to the west, on the other side of the Blue Ridge) petitioned the administrators of the quiet zone for permission to install an improved two-way radio system. That request was summarily denied on the basis that the proposed location of the station was within the quiet zone. The administrators proposed an alternative frequency that proved technically unfeasible; as a result, Rockbridge County's improved system was never installed, and everyone else got the message to not even bother asking. Thus it was that, as of 1969, Nelson and its neighboring counties still depended mainly on landline telephones for their emergency communications.

Whitehead never did accept this as a satisfactory situation. His own picturesque farm sprawled over a wide swath of bottomland surrounded by high hills. His headquarters in Lovingston lay ten miles away, unreachable by line-of-sight radio signals. Rather than petition the FCC for permission, he had taken matters into his own hands and had erected a bootleg antenna at a hilltop facility owned by the local electric co-op. As a return favor—and an example of cooperation between authorities in rural areas where resources are limited—the sheriff's department provided dispatch services for the co-op from midnight to 6:00 a.m. The hilltop transmitter was set up to run on a backup generator in case of a widespread power failure. And, just as Whitehead had suspected, the FCC never did notice this unauthorized encroachment on their quiet zone.

Unfortunately, the necessity of keeping the clandestine radio system out of the sphere of public knowledge also undermined its reliability. When the power failed on the evening of August 19, 1969, the backup generator also failed. It was out of fuel.

There was one other communication system available, and that was the state police radio band. Reception was static ridden, and as far as transmission, Whitehead couldn't get a response to anything he shouted into the mike. It was quickly apparent, however, that a widespread disaster was in progress. A hodgepodge of voices crackled, one atop the other, with urgent reports of flooded streams, destroyed homes, failed bridges, stranded rescue equipment, and missing families. There had apparently been serious devastation along the Rockfish River to the north and even in communities on the west side of the mountains.

Around 3:00 a.m., Whitehead left the rescue squad building. Although it was still raining, Possum Trot had subsided sufficiently that he was able to get as far as the Tye River. That angry stream was now cascading over the bridge, and he watched in awe as it scoured away the approaches. A full-length power pole riding the current struck the submerged bridge rail with such force that it flipped into the air, somersaulted, and landed downstream of the span. In White-head's words, "It acted like it was nothing more than a toothpick."

Stymied again from proceeding further, Whitehead headed home. Headlights were leaving his drive—a pickup truck with three men inside. One was his neighbor and close friend, the rural mailman Tin-ker Bryant. Tinker was seriously distressed. He needed Bill to help him find his wife and three daughters. Their house on Hat Creek, just a few miles upstream of the Whitehead farm, had been completely washed away.

That stream was normally just two to three feet wide near the Bryant home—you could usually step over it—yet the house had been struck by a wall of water that had knocked it off its foundation and carried it downstream. Tinker explained how he'd been holding his wife's hand and leading her upstairs when suddenly he was underwater and struggling to reach the surface. His lungs had felt about to burst, and he considered sucking in a big gulp of water just to get it over with. Then he slammed into a tree and managed to pull himself out of the swirling water, gasping for breath. His pajamas were gone, and he climbed buck naked into the branches. Weirdly, although the rain was still falling in torrents, the flood began to recede, and he noticed a pair of lights below him. It took him awhile to recognize that it was his wife's Volkswagen under the water, and for some strange reason the headlights had come on. When the water dropped further, he climbed down from the tree, staggered to the home of a widowed neighbor, and asked her for some clothes. She gave him her late husband's bathrobe.

Whitehead picked up the two squad members in Roseland and drove north with the pickup truck following to where the road had been impassable earlier. It was now past 4:00 a.m., and the water had dropped, leaving the road heaped with debris that included fragments of Bryant's home: broken furniture, mattresses, smashed appliances, sections of porch railing, and parts of the roof. With flashlights, they set off downstream on foot, shouting the missing women's names: Sarah, Margaret, Patricia, Frances. At daybreak, the sheriff's wife and children began a search for the Bryant family upstream. When the men found a big chunk of the house washed up on a hillside, White-head got a sick feeling in his stomach. He suspected immediately that those four ladies—Tinker's wife and his three teenage daughters—hadn't survived.

That gut feeling would turn out to be tragically correct. The four battered bodies would be found a few days later along the muddy banks of the Tye River, six miles downstream from their former home.

Near the county's southwest border, the raging Tye heaped a colossal pile of debris against the steel piers of a 95-foot railroad trestle. Although the engineers had planned for floods in designing that 680-foot bridge, they had not anticipated such an extreme set of conditions. Nobody knows how deep the water rose behind the debris jam before it toppled the bridge, but it was surely no less than 50 feet and quite possibly higher. With that failure, a massive surge burst downstream, its face a maelstrom of splintered timber, its power comparable to the energy flow over Niagara Falls. Because the bottom of the monster wave was retarded by friction while the top was not, it thundered down the valley not as a wall of water but rather as a cascading breaker, its tumbling mass of debris indiscriminately chewing up everything in its path. This was no run-of-the-mill flood, where victims might struggle to the surface and perhaps have a chance of surviving. Instead, humans and livestock alike were ground downward into the muck of the streambed—where many of their bodies would never be recovered.

The main artery between Charlottesville and Lynchburg, U.S. 29, crossed the Tye River a half mile upstream of the Norfolk Southern's failed railroad trestle. Most folks who traveled this four-lane highway had no particular reason to be in Nelson County; it was just a place they had to pass through to get someplace else. In fact, the entire county had only a single small motel, which had never been known to be filled to capacity. Traffic thinned out considerably after the rain began Tuesday night, but there are always those travelers who for various reasons press on despite deteriorating conditions, and several of them didn't notice—or perhaps didn't believe the sight—that several sections of Highway 29 were submerged. Locals who were helpless to do anything about it saw headlights disappear into the floodwaters. One of the outsiders was lucky enough to have his car swept to a bank, where he climbed out and saved himself. Others, however, were carried down the valley toward the James River. One entire tractor trailer rig disappeared, never to be found.

The washouts, landslides, and piles of jetsam completed the isolation of the southern and western portions of the county. It was impossible for anyone to drive in, nobody could drive out, no equipment or supplies could arrive by rail, the electricity was out, the phone lines were down, and radio communication was dubious. Nor was it possible to go much of anywhere on foot. Every small stream in the region—and there were dozens of them—had overflowed its banks. Many of the creeks had changed course and scoured out new channels. Further hampering foot traffic, most of the low-lying areas were blanketed with a sludge of foul-smelling mud ranging in thickness from two to fifteen feet.

The residents of Lovingston, Nelson County's unincorporated county seat, were dumfounded. How could they be having a flood when the town didn't even have a stream? Rain always ran off the steeply sloped streets rapidly and harmlessly—either to the south, where it drained into Dillard Creek, which in turn emptied into Rucker Run and ultimately the Tye River, or to the north, where it poured into Muddy Creek and ultimately the Rockfish River. Dutch Creek, the single brook on the nearby mountain, ran in the opposite direction, away from Lovingston. Yet by midnight, all of the town's lanes and alleys were inundated by swiftly flowing water. Basements flooded, water rose into some first floors, and the IGA grocery store, the main market in the area, collapsed, its merchandise strewn into the violent current.

A rain-drenched man entered the sheriff's office, his apprehensive wife and two sons trailing close behind. Figuring to spend the night in their newly purchased home, they had driven up from Amherst County and had barely made it this far before the washouts. Was the highway open to the north?

The deputy on duty, semiretired Cecil Davis, recognized the fellow as the new school superintendent, Henry Conner. With classes scheduled to start in a mere two weeks, his main challenge would be to smoothly implement the desegregation plan for the county high school. Hopefully, the storm wasn't an omen of impending disaster on that score. The distracted Conners heard the deputy suggest it would be best if the Conner family spent the night in the sheriff's headquarters.

On a side desk, a state police band radio squawked nonstop with reports of devastation and stranded victims. Several fires were reported, but the fire crews were unable to reach them because of road blockages. Davis answered one phone call after another, all local. There was little he could do other than to counsel the frantic callers to seek high ground and not to try to drive anywhere. He was unable to contact the sheriff, just as Whitehead was unable to reach him. Nor could Davis reach any of the county supervisors. Several nearby residents came in to report emergencies, made a quick assessment of the pandemonium, and left. The entire region was in chaos, homes were collapsing, people were in harm's way, some were even dying, and there wasn't a thing the sheriff's office could do to assist.

Shortly before 3:00 a.m., the building shuddered and they heard a tremendous rumbling. Almost simultaneously, the power and the last of the local phone service failed.

Grabbing a flashlight and a raincoat, Davis scampered across the lawn to the jail and returned with a prisoner who might know how to start the emergency generator. Conner held the flashlight as they fiddled with the machine. No success. Conner began to read aloud from a printed panel. When all else fails . . .

The instructions turned out to be correct, and the generator started. To the educator's credit, he didn't rub it in. After that, they had lights and could again listen to the state police radio transmissions from neighboring counties. The phones, however, remained dead, not just in the sheriff's headquarters but throughout the county. And there was still no radio contact with Sheriff Whitehead.

The deputy didn't bother to return the inmate to the lockup. The detainee took advantage of the opportunity, grabbed Davis's raincoat from the coat rack, and dashed out into the storm. He returned an hour later, the dripping coat stuffed with cartons of cigarettes he'd salvaged from the rubble of the grocery store. He sat down, tore open a pack, lit himself a smoke, and quietly passed the pack around.

Visitors to the Oak Hill Cemetery near the hamlet of Old Myndus seldom fail to notice a recurring date, August 20, 1969, and a recurring surname—Huffman. Sixteen of that clan's disaster victims are buried here; the bodies of six others were never found. Over the knoll from the cemetery, the road veers at a modest but attractive brick home still occupied by Tommy and Adelaide Huffman. Beyond this point, a narrow gravel lane descends into a hollow and parallels Davis Creek for several miles until it rises into the hills. All told, including the twenty-two Huffman casualties, fifty-two people died in that singularly terrible night along a five-mile stretch of the normally benign stream.

There are two principal branches to Davis Creek, neither of which looks particularly menacing unless you're curious enough to wonder about the numerous boulders scattered along their banks. Many of these displaced rocks have dimensions of around five by five by eight feet, which puts their weight at around twenty tons apiece. Some are even larger. Most don't rest in the streambed itself but are nestled on the banks among birches and pines, none of which appear to be older than three decades or so. Whatever was growing along this stream in 1969 was stripped away, and what you see today is postdisaster vegetation.

It takes a great deal of fast-moving water to even budge a twenty-ton boulder. Here, however, hundreds of such rocks were peeled off the mountainsides by falling trees and mud slides, and once in motion, there weren't many obstacles that could stop them until they were good and ready to stop. No avalanche of this magnitude had ever occurred in historical times anywhere in the eastern United States.

What was different about conditions on the night of August 19, 1969? First, the mountain soil, about two feet deep, was already saturated. Normally such soil will soak up a few inches of rain for every foot of depth, but in this event it had no remaining capacity. Second, the storm system stalled over the region for at least six hours. The rain continued to fall at a record rate long after the streams overflowed. Third, the mountains were heavily forested with tall trees whose roots had tenuous grips in the shallow soil. The frost line in these parts extends at least a foot deeper than the soil, and over a period of many centuries numerous tree roots had penetrated the freeze cracks that had developed naturally in the underlying sandstone. When the muddy topsoil gave way, it roared down the slippery slopes, carrying the trees root-first with it. These roots, in turn, acted as giant crowbars that cracked loose huge chunks of the underlying rock.

In spots where only the soil and the trees broke loose, the newly exposed rock outcroppings were found to be well weathered—post hoc evidence that they had been bared to the elements for extended periods in the prehistoric past. Large sections of these picturesque mountains had been denuded before, probably many times, over the course of geological time.

Adelaide and Tommy Huffman found it impossible to sleep in the driving rain. The roof began to leak, and they and their children set pots on the floor to collect the drips. Suddenly they heard a roar like "the sound of a thousand jets" rising from the hollow below.

The electric clock stopped at 2:45 a.m. Tommy picked up the phone to call his brothers, who lived down by the creek, but the line was dead. He was about to head out to warn them in person when a rap came from the door. There stood his brother Russell, Russell's wife, and eight of their nine kids, haggard and drenched. As their car floated off with its engine running and its headlights on, Russell breathlessly explained that they had climbed the slippery hillside to Tommy's place. They'd had to carry the two youngest, ages three and five. Neither had made a peep.

An ominous crash came from the hollow. Adelaide commented on the strange smell wafting in through the open door. "That's fresh earth," Tommy said ominously.

A neighborhood couple showed up, ranting excitedly about collapsing mountains. Tommy knew he couldn't get to his family members across the creek, but there were two more brothers living upstream on this side of the hollow. Leaving the women and children at the house, the men set off in Tommy's pickup truck.

The headlights could barely penetrate the downpour. A few hundred feet into the hollow, they almost ran off the road. They stopped and climbed out. Despite the lack of wind, they were pummeled by twigs and splinters. The echoes of snapping trees and colliding boulders sounded like detonations of dynamite. Given the impossibility of progressing any further, they returned to Tommy's place, praying for the safety of their family members and friends down in the hollow.

Scientists and engineers are fond of quantifying everything they can, and that includes events like rainstorms and hurricanes. Numbers, however, are notoriously difficult to attach to catastrophic phenomena, for the simple reason that the worst pockets of any disaster are precisely those places where unambiguous numerical evidence is least likely to survive. Just as it was difficult to assess Camille's wind speed on the Gulf Coast because of the failure of so many wind instruments, it was equally hard to get a fix on the rainfall in Nelson County because most containers that might function as ex post facto rain gauges had overflowed or were swept away. The resulting numerical picture was a spotty one at best, based on happenstances where a deep open-topped container full of rain had survived. A barrel that had been emptied before the storm, for instance, was found in the back of a pickup truck some eight hours later, filled to a depth of 27 inches. Other buckets and drums were found holding around two feet of water. One report of 31.5 inches could not be independently confirmed because someone began to use that water before any official could measure it.

The rainfall in Huffman's Hollow would be of particular interest to hydrologists, meteorologists, and civil engineers, yet that statistic was especially difficult to ascertain. Investigators would do their best, combining the sketchy evidence with sometimes speculative computations, at least one of which would suggest that as much as 46 inches of rain may have fallen in a six-hour period along an axis that included Davis Creek from its headwater to its confluence with the Rockfish River at Woods Mill.

Regardless of its severity, however, an extreme rainfall alone does not drown people or wreck their homes. It was the secondary effects—the flash flooding and mud slides—that wreaked so much devastation in rural Virginia. Flash floods develop on the upper reaches of mountain streams when an inordinately heavy rain falls on the watershed and overwhelms the ability of the force of gravity to keep the basin drained. Couple this with landslides that temporarily dam such a stream and then collapse, and one has a recipe for a disaster.

The Tye River near Highway 29, for instance, drains a hilly watershed of 92 square miles—roughly equivalent to a rectangle 5 miles wide and 18 miles long. In its previous record flood of 1944, the Tye at this point discharged 9,670 cubic feet of water per second. During the Camille flood, however, this same stream at this same point had an inflow of 80,000 cubic feet per second, or more than eight times its previous record. The river could not accommodate such a huge flow without topping its banks, sweeping away parts of the highway, and toppling the 680-foot-long, 95-foot-high railroad bridge. Six miles upstream at Massies Mill, where the Raines family and many of its neighbors met such tragic fates, the Tye drained a basin of only 66 square miles, and nobody had ever bothered to record its flow even under the worst previous flood conditions. On the night of August 19–20, however, this tiny river had an astonishing flow of 70,000 cubic feet per second as it roared through Massies Mill.

At its confluence with the James River, the Tye, swollen further by its other tributaries and now draining 360 square miles, had a flow rate of 200,000 cubic feet per second, which was more than six times the record flood of 1944 and several times the normal flow of even the James River at Richmond. This tremendous influx from what was normally a tiny stream reportedly caused a section of the James to reverse its flow for half a day.

Meanwhile, the Rockfish River at the north end of Nelson County experienced flash floods of comparable violence; at one point upstream of Woods Mill where its drainage area measured just 96 square miles, the Rockfish peaked at 70,000 cubic feet per second. As for the numerous smaller streams—Davis Creek, Hat Creek, Muddy Creek, and so on—little data was available or retrievable. The only thing known is that these brooks, many with drainage areas measuring only tens of square miles, were transformed into roaring cascades of mud and water, trees and boulders.

Some hydrologists pegged it as a thousand-year flood. Other experts thought that assessment erred on the low side and that a flood like this occurred, on average, only once every ten thousand years.

CHAPTER 13

A COUNTY DIVIDED

Nelson County, August 20, 1969

Filtering through the somber clouds of the east-drifting storm, the morning sun cast a blue-gray pall over the James River valley. Cliff Wood, vice chairman of the county's board of supervisors but foremost a farmer, stepped out his front door and checked his rain gauge. His spread had gotten a hefty soaking during the night—nearly four inches. The river below was a startling sight: swift, red with mud, carpeted with debris. The rising water had already claimed most of the willows lining the bank, and it was now threatening some of Cliff's cornfields.

He drove to a neighboring farm and joined a small group watching helplessly as the river engulfed a pasture and the railroad track. Someone's citizens band radio crackled a distress call: a family was stranded on a rooftop several miles downstream. Cliff gathered up two of his cousins, and they trailered a johnboat as far as they could and then parked on high ground and paddled over the flooded highway. Cliff commented that there seemed to be an awfully lot of water for just four inches of rain. Equally curious, the water swamping the road was surprisingly slack and smooth, considering the fierce current out in mid-river.

They spied the family of three perched on their roof, waving for help. As Cliff and his cousins steered toward them, the man shouted that his mother lived alone a half mile down the road. They ferried the three to the hillside, promised to return shortly, and paddled on.

The older woman wasn't hard to find; she was calling from an upstairs window. Moments later, their boat hit something with a thump. The top of a fence. They shouted to the woman to direct them to the gate.

Unlatching a gate underwater is not a skill many people have bothered to acquire, and Cliff and his cousins were at a loss. They fumbled and tried a variety of ideas to snag the submerged handle, but there was ultimately no way to do it without someone getting wet. Cliff took a big gulp of air, jumped in, and after several attempts to dislodge the latch by Braille, he finally succeeded. It was fortunate that the water near shore was so slow moving. Out in the middle of the river, corncribs and parts of houses were racing past in the frothing current.

As they got the woman into the boat, she broke into tears. She had taken a long time to save up for a color television, which she had finally bought just a few days ago, and now it was destroyed along with her other belongings. And just yesterday, she'd gotten a phone call from a sister who had evacuated from her home on the Mississippi Gulf Coast. They talked about her coming up and staying in Virginia for a while and watching TV together. Who would have ever thought?

As they tried to console the woman, Cliff stood up in the boat and peered downstream. The Rockfish River, which empties into the James, was on a rampage. The Route 626 bridge was gone, and the village of Howardsville, across the Rockfish in neighboring Albemarle County, had essentially been washed away. It was all of that additional water surging into the James that had forced the current on their upstream shore to go slack.

They took the four refugees back to Cliff's brother's general store in Wingina, where the flood was now lapping at the front steps. The approach to the Route 56 bridge was submerged, blocking all traffic to and from neighboring Buckingham County. Sightseers, concerned citizens, and stranded motorists milled about the crossroad, shindeep in water.

The store had two separate phone systems, and although the one serving Nelson County was out, the one connected to points across the James River was ringing with a fusillade of incoming calls. Inquiries about relatives. The electric co-op trying to contact some of its linesmen. Could someone report on the condition of the railroad tracks? Which bridges remained open? As the only person around with any authority, Cliff jumped in his car and headed southwest on the back roads to do some investigating.

He could drive only five miles in that direction. The highway bridge over the Tye River, which had been undergoing an upgrade, had collapsed. All of the construction equipment had vanished. The James River was actually flowing backward, driven by the massive influx from the Tye. Nearby, the village of Norwood was underwater, with most of its structures gone.

Stunned, Cliff drove back to his beloved farm, Arrowhead, and checked in with his wife and daughters. He tried, without success, to use the phone. Not that anyone could do much until the water began to recede, but he did want to make sure that the county's fire and rescue crews, and the sheriff's department, knew about the flooding and the road and bridge washouts. He should explain that there would be no assistance from the far sides of any of the three unruly rivers. It still hadn't occurred to Cliff that the devastation might be more widespread than the flood damage he had seen along the James.

When unfamiliar things happen, it sometimes takes awhile for education and experience to kick in. Cliff Wood was a graduate of Virginia Polytechnic University and a veteran of the Korean War, where he'd served as a military policeman. His family had farmed in the James River valley for four generations, and he knew the region's back roads and mountains like he knew his own face. He was active in local politics and could greet at least a thousand of the locals by name. He'd taken flying lessons for awhile and had given that up only because of the paucity of airstrips in this hilly county. At the age of forty-two, he was lean and vigorous, with a sharp and inquisitive mind.

There had been a bit of academic literature on disasters by that time, but most of it was in obscure journals that never reached people like Cliff. In 1969, even the published experts had not yet noticed that rapid-onset and slow-onset emergencies raise different sets of sociopolitical issues. Social scientists had not yet developed, let alone analyzed, the now-familiar four stages of disaster management: mitigation, preparedness, response, and recovery. And although there was a general awareness that effective disaster management involves four different political units—the federal government, the state government, the local community, and private organizations—few social scientists had ever probed the public policy implications of such fragmentation of emergency authority.

Since the onset of the Cold War in the late 1940s, most of the federal-level thinking about disasters had focused on the prospect of a nuclear attack. Washington's main objective was to maximize the survival rates of those U.S. citizens who were most at risk—in other words, residents of those major cities that were most likely to be targeted in a nuclear missile exchange with the Soviets. A few cities in the North were able to designate their subways and other subterranean enclosures as civil defense shelters, but for most metropolitan areas, timely evacuation was the only practical strategy. Thus did the early civil defense initiative radiate from Washington to the states to the cities, largely ignoring the rural communities. And with few exceptions, this Cold War–era emergency planning had no relationship whatsoever to *natural* disasters.

Virginia was perceived to be at risk from a nuclear attack because of its Norfolk naval base and its proximity to the nation's capital. The Virginia coastline, of course, was also vulnerable to hurricanes. Because of the overlapping risks, the state's legislature had insightfully combined civil defense and disaster management functions into a single state office that answered directly to the governor. This formal arrangement notwithstanding, the main focus continued to be on the prospect of a nuclear war.

In the 1960s, the state began sponsoring emergency management training sessions on a county-by-county basis, delivered by a team of faculty from the University of Virginia. Nelson County's turn came in May 1968. Cliff Wood and the other county supervisors, Bill Whitehead and the rest of the sheriff's department, the school superintendent, and other local leaders all met in Lovingston for a series of sessions that established their roles in a national emergency. Because of the unique structure of Virginia's Office of Civil Defense, those training sessions fortuitously networked Nelson County's public servants with the state officials and agencies whose responsibilities included not just national emergencies but also regional and local disasters.

By his own admission, Cliff Wood hadn't taken the emergency training very seriously. In fact, almost all of the local participants had viewed the operational scenario as somewhat far-fetched: a hypothetical forecast of a nuclear attack on Washington, D.C., which would trigger a mass evacuation of the capital. Members of Congress and other high officials would be whisked by helicopter to a specially constructed bunker at a secret location (in fact, it was beneath a rural resort just over the state border in West Virginia), while hundreds of thousands of common folk would leave the capital by car. One of the main evacuation routes south would be Highway 29, which, except for a bottleneck at Lovingston that was in the process of being alleviated, was a four-lane thoroughfare running the length of Nelson County. Accordingly, the local county officials would be responsible for keeping the traffic moving and tending to any glitches that might arise.

In the drill that culminated their training, the participants were separated into two rooms connected by phone lines while the trainers dealt them hypothetical crises: an overturned truck blocking two lanes of the evacuation route, a radiation-contaminated shipment of food, refugees needing temporary shelter, a report of a rabid dog, a hijacked locomotive, and so on. The task of the trainees was to transmit appropriate communications to one another and to make appropriate decisions. Given the lack of motels in the county, the school superintendent was given the responsibility of providing shelter for evacuees who might become stranded along the way. One of the deputy sheriffs, meanwhile, would coordinate other operations with the fire departments, the health authorities, the coroner, the dogcatcher, and so on. If Cliff and the other participants found that exercise to strain credibility—and apparently they did—the prospect of an actual emergency where the phones didn't work, the power failed, and multiple sections of Highway 29 itself disappeared was beyond the scope of anyone's imagination.

It was going on noon when Cliff decided to head up to Lovingston. Not that he had any specific agenda in mind; the fire and rescue crews were all competent fellows, and they were surely already on their way. Driving to Lovingston was mainly just something to occupy him while the utilities were being restored and the flood had a chance to recede. He'd probably bump into some of the other county supervisors and the sheriff, they'd share information, and maybe they'd commiserate over coffee at Doug's Truck Stop.

The road to Lovingston followed a ridge most of the way, giving Cliff an occasional vista to the northeast, where the runoff drained into the Rockfish River, as well as to the southwest, which was part of the Tye watershed. In both directions, unfamiliar scars pocked the mountains—places where the underlying rock had been denuded of soil and vegetation. Recently fallen trees, their leaves still fresh, criss-crossed the ravines below. The berms of the road were badly eroded in places, and at one spot he had to detour around a washout. Circling down and around the foothills of Naked Mountain, he caught a glimpse of Highway 29. Weirdly, there wasn't a lick of traffic moving on it. Here and there, he could make out small clusters of cars with folks milling around them.

He drove down the winding road into Lovingston. Although the town stood on high sloping ground with no named stream within its borders, most of the buildings along the main street had been flooded and a few had been destroyed. He zigzagged past groups of residents clearing paths and sifting through rubble, acknowledging them but not stopping to chat. Something truly weird had happened last night, and he needed to check on his mother-in-law.

The older woman was fine, but she was in an emotional tizzy about the damage and who was missing and who lost what and who told what to whom. With the promise that he'd return when he had some answers, Cliff drove past the tiny health center, noting with concern that Dr. James Gamble, the town's only doctor, had posted a note that he was on vacation. Cliff swore under his breath. He continued uphill past the single-courtroom courthouse, built in 1810 and still in use, and parked at the annex. He'd taken only a few steps when he ran into Doc Gamble. The haggard-looking physician had postponed his departure because of the storm and had been out and about most of the night—helping rescue a small boy, providing medical attention for a few others, but finding himself helpless to aid the dozens of locals who came to him reporting missing relatives or actual drownings.

Cliff trotted to the sheriff's office and asked about Bill Whitehead's whereabouts. Deputy Davis, still on duty, shrugged; the sheriff had tried to radio in, but the reception was too poor to understand him. One of the department's two communications sets was already out of commission before last night; then this morning, in a well-inten-

tioned but misguided attempt to identify the faulty electronic com-
ponent, a do-gooder started swapping tubes and capacitors between
the two radios. The unfortunate result was that now neither unit
worked. Not to worry, though. Deputy Davis assured Cliff that a
couple of local ham radio operators were on their way there to set up
a communication system.

A communication system, Cliff pointed out, needs functional
equipment on *both* ends.

Davis acknowledged that he realized that. One of the hams would
string up his antenna at headquarters, and the other would head
down to the region near Whitehead's farm and set up on that end.
Just how he'd get there, Davis wasn't quite sure yet. But maybe if
there were a helicopter to take him there. Cliff started to shake his
head in silent disbelief.

The deputy backtracked. He'd neglected to explain that something
may happen before too long. Last night, a landslide at Woods Mill
stranded a state police lieutenant down on Highway 29, and early this
morning that trooper had radioed his superiors in Charlottesville that
Nelson County was in the throes of a disaster. One of the locals had
even taken it on himself to impersonate the chairman of the county's
board of supervisors on the officer's radio since only he could declare
a disaster. Sooner or later, that news was sure to reach the state's
Office of Civil Defense.

As for the actual whereabouts of the other three county supervisors, that was anybody's guess. Given the bridge washouts and road
blockages, it would be a big surprise if any of them showed up in Lovingston in the near future. If there were decisions to be made, Deputy
Davis suggested, it would be up to Cliff to make them.

Cliff took a deep breath. He well knew that you can't make decisions without relevant information, and he certainly wasn't going to
get any information from the outside about the extent and the degree
of the damage here inside the county. He grabbed the key to the civil
defense jeep, invited one of the locals to join him, and set out to
inspect firsthand as much as he could.

Five miles north of Lovingston and a mile off Highway 29, Tommy
and Russell Huffman had been struggling since dawn to make their
way on foot along the south bank of Davis Creek. Russell's house was

gone; they'd established that fact by flashlight as soon as the rain had quit. Fortunately, all but one of Russell's immediate family had escaped to Tommy's place. The big unknowns were the status of their five other brothers, one of Russell's sons, their mother, a couple of uncles, dozens of cousins and in-laws, and a number of neighbors who for all practical purposes were also family. All had homes and small farms in this narrow valley.

Nothing resembling a road remained. For hours they struggled through a mass of uprooted trees, mud, and boulders—some the size of a car. Tree trunks up to four feet in diameter had been snapped, and in some spots on the hillside the debris rose sixty feet above the creek bed. Tangled in the rubble were fence posts, sections of barn siding, a piece of a porch spindle here, a smashed hog trough there. Across the creek, which was still roaring but gradually returning to its banks, they saw a section of roof sitting in a pasture. They recognized it as part of their brother Robert's house.

Numerous spring branches run laterally into Davis Creek, and cradling each of those little brooks is a narrow hollow that rises steeply to the surrounding hills. Some of their kin had built houses or set up mobile homes in these coves, choosing the ones that seemed to be safely above the creek. The prospect that a tiny spring branch could threaten a home, let alone destroy one, was beyond the pale of anyone's imagination. Yet overnight, thousands of tons of trees, stones, and mud had avalanched down the sides of most of these coves.

Each footstep was precarious, and the going got no easier. They saw no bodies except for a few farm animals, and not even many of them. Yet who could say what awful news lay buried beneath all of the wreckage and mud?

A black neighbor, Foster Jackson, ran down the hillside on the opposite bank, waving frantically. His initial words were lost in the roar of the water. To get above the din, they climbed higher as Foster did the same on his side of the hollow. The message turned out to be far from encouraging. "They're gone. They're all gone!"

They hiked back to the former site of Russell's home. It was still impossible to cross the swollen creek. Across the valley, they glimpsed their uncle Sam, his wife, and their son and their dog, sitting exhausted on the grassy slope. They had climbed farther up the

One of many farm buildings destroyed by the great flood in Virginia. *(Courtesy M. E. "Ed" Tinsley.)*

mountain during the flood and spent a dreadful night in the woods. Tommy waved. Sam stood and yelled that their home was destroyed. Sam screamed to Tommy and Russell that their mother's house was gone too and that she was injured and needed help.

The brothers returned to Tommy's place and told their wives the good news about Uncle Sam being okay. As for the bad news—well, they still weren't ready to admit to any bad news beyond Russell's home being gone. They needed to head out to Highway 29 to get some help.

They tossed two chainsaws and a can of gas in the bed of Tommy's pickup and set off, stopping to clear the lane where they needed to, swerving around obstacles where they could. That gravel road ran on high ground most of the way, and the obstructions were minor compared to the mess they'd encountered down along Davis Creek.

A state highway maintenance truck met them from the opposite direction. The driver had been dispatched to the Lovingston District the night before to check out a report of a "small" landslide, and when all hell broke loose he got stranded. He had a two-way radio.

They led him back to Tommy's home. As the two trucks pulled into the drive, another frazzled neighbor, a black fellow from up the hill, dashed to them with the news that several families were trapped and there was no way to get to them. The state worker peered down into the devastated hollow and then radioed his headquarters in the next county. Send help, he told his boss. Lots of help. Quickly.

Sheriff Bill Whitehead remained trapped at the far end of the county inside a circle less than five miles across, all prospects of overland egress blocked by landslides, bridge failures, and washouts. In nearby Roseland, which Whitehead could look down on but couldn't get to, all twenty homes, the post office, and the two businesses had been swept away. Five days later and 240 miles downstream near Hampton, Virginia, a boat in Chesapeake Bay would snag the Roseland post office sign in a fishing net.

As Whitehead organized and dispatched the first search parties, he made repeated vain attempts to raise a radio response from his Lovingston office. In contrast to that terminally silent frequency, he found the state police radio network clogged with a cacophony of frenzied voices transmitting simultaneously with total disregard for the protocol of waiting for an instant of silence before talking. Yet through that hubbub—a barely decipherable phrase here, a word or two there—he got the picture that this was more than an isolated event. All of Nelson County seemed to be in trouble, as were many portions of the neighboring counties.

Figuring that his message had to be at least as urgent as anyone else's, Whitehead bellowed into his mike that the Massies Mill District was isolated and that they urgently needed helicopters. Nobody acknowledged hearing him. Over the next several hours, he transmitted a series of variations on that plea on the state police band, asking the recipient—whoever that might be—to relay the request to the civil defense authorities. Helicopters. They needed helicopters.

In the meantime, he was not a man to sit and twiddle his thumbs. He drove one back road after another, reversing when his progress was blocked, zigzagging around obstacles where possible, and got close enough to Massies Mill that he was able to climb down an embankment and wade through shallow water to get to the bridge at the lower end of the village. The town's agricultural supply store— the one managed by Carl Raines—rested squarely on the span, blocking both lanes. Crouching under an overhanging section of the wrecked building, he made his way to the opposite bank, where he again had to slosh through water to reach the devastated community.

The sight was unfathomable. Many of the forty homes were shattered beyond recognition, others were totally gone, and a thick layer of debris-strewn ooze blanketed the whole region. There were strange scars on the nearby mountainsides, as if an angry rain devil had reached out and clawed them down to their bones. A distinctive smell permeated the air. It was the pungent odor of what the mountain folk call "woods dirt"—earth that has been stripped from the hills after clinging there undisturbed for centuries.

As the first visual semblance of authority to arrive in the chaos, Whitehead was immediately swarmed by dozens of residents. Although at this point the Tye had receded from twenty feet above normal to just seven or eight feet, the roar was still continuous, and verbal communication required shouting almost directly in the recipient's ear. Several rescues were in progress, using ropes and/or human chains. A few had already been successful.

The Raines boys, Warren and Carl Jr., had been recovered from separate trees. Ragged and muddy, they dashed up to the sheriff and asked if he'd seen their parents or siblings. Whitehead would later remark that he'd never felt as helpless as he did at that moment, ostensibly being the guy in charge yet having absolutely no information to give the two teenagers. The best he could do was to assure them that he'd let them know immediately as soon as he received any news.

The most poignant sight, however, was that of the Raineses' white clapboard home. Although the lower floor had been flooded and its windows shattered, the house remained structurally intact and firmly on its foundation, its second floor unscathed. The ruins of another house from farther upstream had lodged just north of the residence and had diverted the swiftest part of the raging current. Had the Raineses stayed put on their second floor rather than evacuating, they would all have survived.

The disappointed boys left the sheriff, sloshed through the mud to their home, and entered the silt-covered first floor. The kitchen was wrecked; the appliances had floated and had smashed everything in that room. Flood-borne debris had destroyed the plaster in the other downstairs rooms as well. They heard a knocking sound upstairs. They shouted for their mother, their father, brother Sandy, sisters Ginger and Johanna, but there was no answer, just the knocking.

Many homes suffered major damage or total destruction. *(Courtesy M. E. "Ed" Tinsley.)*

They climbed the stairs, calling. The sound was coming from one of the bedrooms. They pushed open the door. In Carl's bed, his tongue hanging out and his fur matted with mud, was their black Labrador retriever, Bo, his wagging tail thumping against the headboard.

They didn't admit their worst fears to each other and maybe not even to themselves. After all, they had survived and the family dog had made it, so surely they should stifle any doubt that the rest of the family was okay. In fact, Mom and Dad might even be out looking for *them*. Warren and Carl changed out of their soggy clothes and set out to resume their search, inviting Bo to go with them. The Lab would have none of it; he insisted on staying right where he was.

Through the rest of the morning, a flurry of reports reached Massies Mill by foot courier, telling of stranded victims, dead bodies hanging in trees, and dozens of rescue attempts—some successful, some not. Bill Whitehead tried to evaluate the credibility of each story, dispatching members of a growing group of local volunteers to assist where they would be most likely to make a difference.

Late in the morning, he got word from a linesman that the emergency generator at the repeater transmitter had been out of fuel last night but that now it was gassed up. He again fired up his sheriff's radio. To his chagrin, he still couldn't raise a response from headquarters. In the interim of failed radio communications, the radio equipment at the Lovingston end had suffered its unfortunate demise.

Around noon, a small two-seat helicopter from the phone company flew over the area, and Whitehead flagged it down. It landed near the elementary school, its skids sinking into the muck. The pilot declined to take a man as large as the sheriff up in that small craft, but he did agree to accept two young boys as guides to check out a report of two trapped girls. The report turned out to be accurate but untimely, insofar as those particular girls had already been rescued from the ground.

As that mission was under way, a U.S. Marines helicopter thundered up the valley. The craft was already low on fuel after conducting several evacuations in the neighboring county, but at the request of the Augusta County sheriff, who had heard Whitehead's pleas on the radio, the crew was making a quick aerial inspection. Whitehead signaled the pilot to land, and they loaded up a dozen people who needed medical attention.

Unwilling to risk the loss of the badly needed chopper, Whitehead tagged along on the flight to Lynchburg General Hospital, followed by a refueling stop at the Lynchburg airport. Only then did he learn that the helicopter would not be placed under authority of his sheriff's department. A distant U.S. Marine officer was making all decisions about the chopper's use, and the crew's orders were to return to a site on Davis Creek where a man and a woman had been spotted trapped in a debris pile.

Cliff Wood was still struggling to assess the situation from the ground. At Freshwater Cove Creek, just past the high school and less than four miles south of Lovingston, Highway 29 was washed out and a smashed car lay down in the creek bed. The word was that the driver—someone from outside the county—had drowned. The sheriff's office had been contacted, or so the bystanders said.

Cliff wheeled the jeep around to survey the situation north of Lovingston. He passed a house resting on the median and then, a few miles farther, the wreckages of other homes. Five miles north of the county seat, a highway crew was struggling to clear a huge mountain of debris, using equipment that fortuitously had not yet been trucked away after the recent completion of the Highway 29 bypass. Although that initiative was well and good, it was obvious that it would take more than a few bulldozers to open up the road for any significant distance. North of the obstruction, Cliff could see that a long section of the highway was heavily damaged, with segments of the bridge's upstream lanes missing.

He got out and walked around. A heap of rubble five stories high and covering several acres was piled up at the juncture of Davis Creek and the Rockfish River at Woods Mill. Locals were wandering around half-dazed, exchanging rumors of one family tragedy after another and raising as many questions as answers about the scores of folks who were missing. They told Cliff about the devastation up Huffman's Hollow, of huge boulders and whole trees cascading down Davis Creek, sweeping away cars and bashing homes to smithereens.

Cliff's mind started growing numb. What chance might there be that any of the missing people up the hollows were still alive? If they were, what was the chance of locating and rescuing them? And if rescued, what was the chance of treating the injured in time to save their lives? This situation was a whole lot different than the hypothetical scenarios they'd practiced during last year's civil defense training. At that time, the operating assumption was that you could phone people and the roads would be open to send the sheriff or his deputies where they were needed. Hell, Cliff didn't even know if the sheriff was still alive. Meanwhile, he was barely aware of the distant sound of a helicopter.

Cliff trekked back toward the jeep, his mind reeling. His corn crop didn't seem so important anymore.

A stranger walked up to him as if they knew each other. He was a large man with slick dark hair, built like a NFL lineman and wearing a clean white shirt. He introduced himself as Jim Tribble from the state's Office of Civil Defense. Cliff identified himself as one of the four county supervisors. Jim asked if any of the other supervisors were around.

"No," said Cliff, "there's no way any of them can *get* here."

"Well, then," Jim asked, "would *you* say this is a disaster?"

Cliff bit his tongue to stifle an expletive. He replied as calmly and politely as possible that, yes, indeed this was a disaster.

Big Jim nodded. "That's what I needed to hear." There could be no assistance from the state unless the local authorities first declared a disaster, and County Supervisor Cliff Wood had just done so. From that point on, the wheels could turn.

In fact, Jim Tribble had already conducted an aerial survey of much of the county and had even landed at the rescue squad building in the Massies Mill District, unsuccessfully trying to locate the sheriff. What he'd seen with his own eyes already gave him more than enough information to convince the governor to declare Nelson County a disaster area. Not a single stream remained within its banks, debris was strewn everywhere, large homes lay toppled on their sides, smaller homes lay scattered in fragments. Every bridge in the county was either gone or damaged. Not a single main road was passable for more than a few miles, and many of the unimproved roads had vanished completely beneath landslides. Cliff was stunned by Big Jim's bird's-eye description.

They drove back to the recently completed Lovingston bypass. Several two-place Bell copters were buzzing overhead, and one Marine chopper had come and gone, along with Sheriff Whitehead and some chainsaws and other tools borrowed from an electric co-op service truck. "You need to set up an airstrip," Big Jim advised Cliff. Helicopters can land in lots of places, but fixed-wing aircraft need a straight and level landing run of three to four thousand feet. Cliff, who used to fly a bit himself, recognized that the issue was power lines; they aren't always visible to pilots from a distance.

Before leaving to file his official report, Jim asked Cliff what else the county needed. Cliff still had no way to give a specific answer other than that they needed communications, food, and more helicopters.

As Jim took off, Cliff mobilized several bystanders to park school buses and an insecticide truck as barricades at the ends of a mile-long segment of the bypass, while he sent others to tie bedsheets to the power line at the north end of the strip. They were the first directives of hundreds that Cliff would give in the coming weeks. He watched half-stunned as everyone sprang into action without a question.

That task was barely complete when two army Hueys landed, the lead pilot having gotten a call that twenty people were stranded at Variety Mills—which wasn't on any maps. As Cliff climbed into one of the choppers to navigate, he noticed several ham radio operators setting up equipment on their tailgates. A short time later, a voice crackled from the helicopter's radio with an instruction obviously intended for some other aircraft: "Land in the southbound lane." One of the hams had unilaterally set himself up as the air traffic controller at what would temporarily come to be known as "Nelson International."

Some men seek power, some men don't give a darn about power, and some take awhile to adjust when power is thrust upon them. Cliff smiled at the ham's boldness. The realization that Cliff himself was the guy in charge had yet to sink in.

On a meadow near Davis Creek, Sheriff Whitehead and the crew hopped out of the helicopter and toted the chainsaws to the creek. There they found an elderly man and a woman trapped back-to-back in a mound of debris. The woman's status had been questionable from the beginning, but the man had still been talking when the report came in. Now both were limp. Whitehead climbed over the rubble and confirmed that neither victim had a pulse. The pilot picked up another emergency call. Whitehead passed the chainsaws to a group of locals, promising that the chopper would return to pick up the bodies. Some folks' lives might still be saved.

A few miles downstream, they circled Tommy Huffman's home. Tommy stood in the yard and gestured toward the opposite side of the creek. As the chopper landed there, Sam ran up. Three of the kids had been found alive, he told them excitedly. It was the dog that did it, by running back and forth between Sam and the ruins of Ma Huffman's home until Sam and his wife went to explore. There they found the toddlers under a section of a wrecked roof gable, terrified but unscathed.

Ma Huffman, however, had a leg laceration that went to the bone.

And her son Jesse had a series of deep wounds on his back. White-head didn't bother making records of who all was okay, who was injured, and who was still missing; he helped the old woman and her son into the helicopter, and they took off to the hospital in Lynch-burg.

In the turmoil of having twenty-two deaths in the same family in one night, many of which took awhile to confirm, Jesse and his mother temporarily got lost. The surviving Huffmans erroneously assumed that Jesse and Mama had been airlifted to Charlottesville, and that's where they went to look for them. It was several days before they got word that the family was recovering in Lynchburg.

Cliff Wood and Bill Whitehead, county supervisor and county sheriff, spent that entire Wednesday afternoon in separate helicopters. At dusk, Cliff returned to his home near Wingina, forced to hike the last mile through pastures because the road had become impassible during the day. Meanwhile, the Marine helicopter crew dropped the sheriff off at the rescue station near his own farm on Hat Creek.

Both men's minds were racing, generating more questions than answers about what to do next. They still couldn't communicate with each other by phone or radio, and their brains didn't always run on perfectly parallel tracks. And although their values were the same, their immediate issues were not. Cliff needed to coordinate the county's transportation and communication hub, which meant dealing with outsiders. Bill Whitehead needed to address the needs and unknowns of the totally cut-off Massies Mill District, while still performing his regular duties as sheriff. He had also committed himself to spending a second day helping the Marine helicopter pilot find his way around the county.

Whitehead stayed up late, digesting reports about things that had happened locally in the last twenty-four hours. Due to the heroic efforts of several linesmen, a cluster of three buildings now had power: the small rescue squad building, the funeral home, and Bethlehem Methodist Church. The church was the obvious place to set up an emergency center, and in fact some resourceful women had already been serving food there since early afternoon. A civil authority, however, can't simply move his operations into a church without an invitation, and the minister and his wife were stranded by the flood and couldn't be reached by phone. Whitehead sent word of his dilemma, by courier, to the congregation's leaders. A few hours later, he received return confirmation that it was okay to use the church as the center of emergency operations and that enough parishioners had already volunteered to staff it.

Meanwhile, there were displaced families in need of temporary shelter. There was the issue of distributing food from those folks who had a surplus to those who had nothing to eat. Although fresh springs were gushing everywhere up on the hillsides, safe drinking water was a problem in the low areas, where manure and fertilizer-laden floodwater had entered cisterns and wells. And then there was that depressing pile of messages about bodies that had been found and folks who were still missing. An important function of the emergency center would be to serve as a clearinghouse for inquiries and notices, plus the grimmer announcements.

Roads were also a major issue, for there is always a limit to how long any isolated region can survive on its own. Food, supplies, heavy equipment, and fuel would need to be brought in fairly soon, and all of that couldn't be done by helicopter. Yes, a small bulldozer belonging to the electric co-op had started filling in the smaller washouts and clearing the lesser landslides, and privately owned chainsaws had been buzzing all day. Those early efforts, however, were piecemeal and uncoordinated. Although the sheriff wasn't about to start directing the road-repair volunteers himself, he did want to make sure that everyone knew how they fit into the big picture. He studied his county map, marked the points of the known landslides and washouts, and made his best guesses about where the volunteers might open up detours around the bridge failures and obstructions on Route 151 and the west end of Route 56, which ran up over the Blue Ridge and down the other side of the mountains into Rockbridge County. The first thing tomorrow, he'd hang the map on a wall of the emergency center so the crew chiefs could add and subtract information as the days wore on.

And then, exhausted, Bill Whitehead shut his eyes to catch a few hours' sleep.

CHAPTER 14

RECONNECTING

A hurricane, in and of itself, may or may not be a disaster. The same goes for avalanches, volcanic eruptions, earthquakes, or blizzards. Mother Nature often unleashes violence that does not undermine society (a flash flood in an uninhabited canyon, for instance, or a typhoon over the ocean). Even an airplane crash that claims dozens of lives can more properly be described as an accident than a disaster. An essential characteristic of a disaster is its social disruption—the inability of the affected society to continue on as before.

The first thing that usually fails is communications. When Camille threatened Louisiana and Mississippi—which were both slow-onset disasters—the half day's warning gave police and fire departments time to test emergency generators, change batteries in radios, alert emergency workers about rendezvous sites, and so on. In the rapid-onset disaster in Virginia, no warning or no preparation was possible. In all three places, however, the ability of relevant social units to perform was undermined almost immediately by the arrival of the storm. Several days would pass before communications would be reestablished among the critical public agencies, and many of the victims would be lucky to have their phone service restored even a month after the disruption.

Even as communication is failing, its importance is increasing. Word needs to reach the right people about a wide variety of unanticipated collateral effects of the crisis: fires, escaped animals, victims trapped in strange places, chemical spills—and the list of possible bad surprises grows with the severity of the event. The more urgent the need for communications, however, the more likely it is that the communications infrastructure itself has been wrecked.

The second thing disrupted is transportation. Louisiana and Mississippi had the benefit of enough advance warning to organize an evacuation, and their local leaders knew, at least in general, what problems to expect in keeping traffic flowing. In Virginia, there was no evacuation because there was no warning. Road and bridge failures during the storm trapped and isolated survivors in all three regions, but it was in rural Virginia that the destruction of the transportation infrastructure had the most severe social impacts. Louisiana and Mississippi were accessible by sea, so heavy equipment and supplies could eventually arrive that way; Mississippi had several airports that were relatively undamaged; and most of the road blockages near the Gulf Coast were due to trees and other debris, which, albeit with difficulty, could at least be bulldozed aside. In Nelson County and neighboring parts of Virginia, however, complete infrastructural failures were more widespread and more severe, with more than two hundred miles of roads and ninety-five bridges washed out or heavily damaged. Without functional roads, the delivery of essentials such as health care and safe water was seriously hampered.

All three regions had to cope with additional hazards in the aftermath of the storm. The coastal victims were plagued by mosquitoes, snakes, and fire ants. Sanitation was a major issue (as Jackson Balch indelicately testified at the hearings, what goes in one end eventually comes out the other). Failures of the levees in Louisiana left victims vulnerable to a follow-up flood should another hurricane happen along. Nelson County lay exposed to a similar risk: with its streams clogged with debris and many silted up and wandering out of their natural channels, even a modest follow-up rainstorm might cause serious flooding. Compounding matters, all three disaster zones had chemical hazards that needed to be addressed: leaking drums of toxic fertilizers, petrochemical tanks, and so on. Although water was everywhere, virtually none of it was safe to drink.

The most emotionally difficult task in the aftermath of any disaster is the recovery and identification of the dead. When, in normal times, authorities are notified of an accidental death or a missing person, search and recovery responsibilities fall under the official duties of institutionalized police and fire units. In a disaster, however, where hundreds of people die simultaneously and collateral infrastructure failures confound the searches, there is no way to accomplish the grim task of hunting for bodies without teams of volunteers. Formal administrative structures give way to ad hoc groups whose coordination presents novel problems and whose effectiveness can be highly variable.

In Mississippi, the availability of the Seabee base was a big advantage, as was the fortuitous presence of the hundreds of young airmen at Keesler Air Force Base. Working together under minimal supervision, these servicemen were able to sift through the rubble and recover most of the bodies in the first few days after the disaster. Anyone still missing along the Gulf Coast after a week or so had probably been washed out to sea and would never be found.

In Nelson County, recovering the dead was considerably more challenging. There, the affected areas were less accessible, and many of the bodies had been swept miles downstream. Relatively few of the bodies were recovered immediately, and the organized searches—assisted by hundreds of volunteers from neighboring counties and Mennonites from northwest Virginia, Pennsylvania, and Ontario—would continue for a full four weeks. One corpse would be discovered in December, four months after the disaster. For many Nelson County survivors, this extended search had the effect of dragging out the psychological horror of the response period and delaying its closure. It's hard to think about rebuilding your home when you still don't know what happened to your family.

Upheaval and readjustment, both personal and social, are the hallmarks of disasters. Self-sufficiency, already emasculated by the sudden destruction, becomes a chimera as outside authorities undermine old assumptions about who is really in control. Recovery, paradoxically, becomes a time for action yet a time for caution. This quandary of ego versus need has probably never been more disconcerting than it was for the residents of those three conservative regions so cruelly hammered by Camille.

Plaquemines Parish

Although less paranoid than the late Judge Perez regarding federal and state assistance, Luke Petrovich was still wary of outsiders. He prohibited news reporters and teams of relief volunteers from entering the devastated peninsula, and insurance adjustors had to register and be issued passes. The twenty-mile-long lagoon swamping the south end of the parish, however, wasn't about to disappear on its own. The Army Corps of Engineers couldn't start too soon for him, and he likewise welcomed the Louisiana National Guard to help with the levee repairs. Meanwhile, the evacuees were bused in just far enough to confirm for themselves how bad things were, but then they were returned to their shelters at the north end of the parish and counseled about how to seek temporary housing up there.

The engineers brought in more than a hundred farm tractors by boat, lined them up on the levees, connected them to irrigation pumps, and ran them night and day for three weeks until the flood-water was returned to the river. Next would come the scrapping of grounded barges and the removal of wrecked boats and massive quantities of debris. Structures and equipment that might still be salvageable needed to be carefully preserved. Electric and phone lines would need to be strung anew. Just completing the preliminaries was a long, slow process.

In the midst of dealing with all this, Luke was out with a crew one day when he saw a young black and white bull calf trying to maintain its balance on a piece of floating debris, a look of fear and utter helplessness in the animal's eyes. The vulnerability Luke saw in the calf's eyes tapped deep into his own feelings of helplessness during the long hours spent in the sewage treatment plant. Although surrounded by the complete devastation of lower Plaquemines Parish that required his immediate attention, he asked a helper to wade through the four feet of standing water to rescue the animal. They succeeded in getting the calf up on the levee, only to have it die anyway.

Luke set up the parish's temporary disaster headquarters in an old riverboat moored at the levee. Nothing happened automatically; decisions had to be made about everything from where to deliver fuel to who should get phone service and power restored first to what to do about surviving livestock and lost dogs that were starving and becoming vicious. Fire ants and snakes were a huge problem, and several workers had to be hospitalized for bites or stings. Debris piles couldn't be burned without painstaking prior inspection for fear of burning a human corpse. Coordinating everything kept Luke busy nonstop, not for a few weeks or months but for the next few years.

All politicians have their gadflies, and Luke was no exception.

Before the disaster, the elderly Mrs. Williams (not her real name) had been phoning him and the sheriff almost daily at 5:30 a.m., telling them how to run the parish. She'd then spend the rest of her day informing various friends about the earful she'd given the authorities that morning. If it was possible for the disaster to have any upside, Luke quipped to the sheriff, at least Camille had stopped those phone calls when she knocked out the phone lines.

On word that the monumental pumping job was finished at last, Luke did a flyover inspection in a helicopter. He'd done flyovers before, but this was the first time he could inspect the site of his own home, and he told the pilot to set the bird down. They got out and silently tramped around in the mud. Virtually nothing was left. His mother's dishes would eventually be found almost a half mile away, and his own bedroom three miles away. But there lay the telephone—undamaged. The pilot picked it up, held the muddy receiver out to Luke, and said, "Mrs. Williams calling." Both men burst out laughing; it was the first time Luke had been able to release any tension since the whole nightmare began.

Mississippi

Ben Duckworth remembers nothing of the drive back to Jackson; he suspects he slept the whole way. Somehow his mother was prepared for their arrival, and the table was piled high with food even though it was after midnight. Josephine told Ben about her prayers, her phone calls, her unwavering faith that everything would turn out fine. She peppered him with questions. Ben had little to say; his mind was still a blur of confusion. The only thing he knew for sure was that it was great to be loved.

Even prior to the disaster, Ben was prone to drifting into melancholy when he thought about the failure of his marriage and the geographical remoteness of his young daughter in Texas. Now, compliments of the hurricane, a dozen people he cared about and had grown close to were dead. He couldn't help feeling that there must have been something else he could have done—maybe something quite minor—that would have tipped the balance from death to survival for some or all of them. As he lay in bed at his parents' home, staring at the ceiling, he couldn't rid himself of the feeling that he'd

One of hundreds of puzzled pets. *(USACE, New Orleans District.)*

let everyone down—his friends, his daughter, even his parents. Now all of his personal possessions were gone. He no longer even had a job. What was his role in life? Where did he belong?

Languishing in Jackson, Ben lay outside the systems of mutual support that arose naturally among the survivors in the stricken neighborhoods. Although his family and hometown friends were loving and supportive, they had not been part of his personally harrowing experience, and despite their best efforts they would never fully connect with the feelings that churned inside him. For a few days, bed seemed to be the only place where he could begin to grapple with his confused thoughts, struggling to make some transformation from what had passed to what he would do next. In such instances, people often find security in latching onto tidbits of prior normality. Ben began thinking about his car. Maybe, it occurred to him, just maybe, there was a remote possibility that his car had survived.

He dressed, kissed his mother, and went to the Chevrolet dealer in Jackson where he'd purchased the new Impala a few months earlier. The sales manager looked up the serial number and made him a new set of keys. With his parents' financial assistance, he bought himself a new pair of glasses, a watch, a wallet, and some clothing. Most of the churches in Jackson were collecting relief provisions and trucking them to the coast, and Ben hitched a ride with one of the groups heading off to deliver aid.

He'd made this drive dozens of times in the past, and a whole series of inconsequential landmarks had been subconsciously stored in the recesses of his memory. A fruit market here, a dairy silo there, a car lot, a gas station. Now fallen limbs and piles of sawed logs intruded on the landscape. Bright blue tarps covered damaged roofs, rough boards barred broken windows. As the miles wore on, the destruction escalated to vehicles flattened, barns toppled, and houses so crippled that not enough structure remained to bother to tarp them.

Entering Gulfport, he ceased to notice many of the old familiar landmarks. No longer were the debris piles intermittent; they now merged into continuous heaps on both sides of the highway. Chain-saws buzzed as high-end loaders cleared paths down the side streets. Human activity was everywhere, and everyone seemed to be part of some team: National Guard troops patrolling in open jeeps loaded with fuel cans and water jugs, women serving food from makeshift tables of splintered planks and sawhorses, utility workers replacing downed power poles, children piling up brush to be loaded or burned. Even the stray dogs sniffing around in the rubble were doing so in groups of two or three.

They drove on to Pass Christian, where the driver dropped him off at the Catholic church. The main door was ajar, and he walked in. The building's rear wall was gone. Poking from a pile of bricks in the alley behind the ruined sacristy, he glimpsed the remains of his car, its windshield smashed and its roof crushed. The high water mark testified that the vehicle had been flooded.

He fished the key from his pocket and made his way through the scattered bricks to the driver's door, the only one accessible. The side window was still intact, pine needles embedded in it and some even piercing through it, compliments of the brutal winds. As he yanked open the door, a voice shouted from behind him: "Mister, that key better fit!"

He turned to find himself looking down the barrel of a shotgun, a middle-aged man's finger on the trigger. The fellow was alone, the only person Ben had seen so far who wasn't part of some team. Ben had a fleeting thought of the irony of having survived the catastrophe only to be killed for trying to recover his own car. "It's mine," he said.

He climbed inside, hunching under the crushed roof, and found to his relief that the key slid smoothly into the ignition. He twisted it. Surprisingly, the engine cranked. More surprising, it coughed, sputtered, and then started. He climbed back out, spent a few moments gazing in astonishment at the wheezing wreck, and then started to clear away the bricks. The man with the gun moved on without an apology.

Somehow, successfully starting that battered car triggered a kind of rebirth. He'd had a life before Camille, and now he knew in his gut that he'd have a life after Camille. Crouching beneath the crushed roof, he drove to the home of friends in Ocean Springs—to enjoy their surprise, their attention, and their beer, in celebration of the miraculous survival of Ben Duckworth and his flattened Chevy Impala.

He would never again live on the coast. He drove the smashed Chevy back to Jackson and took a job at a bank. Three days into that new career, he met a lovely female coworker. Barely a year later, they were married.

Nelson County, Day 2

The calendar dates no longer mattered. Throughout the county, Wednesday, August 20, became Day 1, and the counting went on from there. Warren and Carl Raines spent that night at the home of one of their dad's customers. Nobody had said so explicitly, but the unmistakable message in the eyes of everyone they met, including their overnight hosts, was that, if their parents hadn't shown up by now, they weren't likely to be found alive.

Giving up hope, of course, doesn't come easily. The boys returned to Massies Mill in the morning. The store their dad had managed, the one that had washed up on the bridge, was being demolished by a team of volunteers with chainsaws and crowbars. Other men were retrieving scattered drums of agricultural chemicals from the muck of the riverbanks. Warren and Carl found their house still empty except for Bo. They fed him a sandwich from Carl's pocket and went back outside. Gingerly, Bo followed.

A state trooper had just arrived via the back roads from across the Blue Ridge, and he yelled to the boys. Their hopes surged, but only momentarily. The officer, Ed Tinsley, merely wanted to ask them what they were doing inside the empty house. Warren explained that they lived there. As for the whereabouts of their parents, they couldn't say; they had hoped the trooper knew something.

After quizzing them for a few minutes, Tinsley unclipped his walkie-talkie and stepped away to make a series of calls. When he signed off, he told the boys he'd take them to their grandparents in the next county. They'd leave as soon as the bridge was cleared. Meanwhile, the boys needed to find someone to look after Bo.

Warren took exception. He and Carl needed to stay to make funeral arrangements, he explained. More than thirty years later, that comment would still haunt Tinsley. A fourteen-year-old boy worrying about arranging the funerals of his parents and several siblings.

Tinsley, however, insisted. He assured Warren and Carl that they'd be informed immediately as soon as there was any news to report. When one lane of the Massies Mill span was open, he drove them to the Route 151 bridge on the Piney River. Although this tributary of the Tye had also flooded, it had done so with less violence than most of the other local streams, and a group of locals had already cleared the approaches sufficiently to allow the patrol car to squeeze through. They parked on the berm on the Amherst County side. As they waited, several pickup trucks passed in the opposite direction, loaded with chainsaws, gas cans, tents, and tools.

A sedan pulled up and three people spilled out—Warren and Carl's grandfather, their oldest sister, and one of their mother's uncles. Their sister asked where everyone else was.

"We don't know," the boys said. "They're gone."

Their great-uncle, a decorated World War II veteran, sat on the guard rail and cried like a baby.

A state police monoplane arrived in Lovingston early Thursday morning, and the pilot took County Supervisor Cliff Wood up to survey the damage. The town of Schuyler was hard-hit, but at least it was partially intact and, most important, remained accessible by road from neighboring Albemarle County. A few other hamlets near the eastern county line could also get help from the outside. Cliff was mainly concerned about the areas that were cut off—about making sure that none of them would be ignored. He directed the pilot to zigzag over one square mile after another, to circle around clusters of homes, and to follow the courses of misbehaving streams and washed-out roads as he carefully committed it all to memory. Meanwhile, he couldn't help but notice that the helicopter traffic had increased.

When he landed around noon, Cliff was met by an official from the Office of Civil Defense who was fretting about the confusion around the airstrip. All of those whirlybirds were flying around with nobody keeping track of where they were going, who was on them, who was being evacuated, and so forth. There was no rhyme or reason to any of the flight patterns, and nobody was assuring that the same spots weren't being searched over and again while other places might be getting missed. Although the official had no authority to order a county supervisor around, he suggested that Cliff set up card tables alongside the landing strips and appoint someone to log in each flight and maintain records of where the crews had gone and what they'd done. Cliff accepted the scolding silently. He hadn't yet been thinking about logistical issues; so far, he'd been too busy just trying to ascertain how widespread the destruction was and which areas needed attention first.

Able and cooperative volunteers weren't hard to come by. As the tables were being set up, several men drove off to find a large tent. Another pair scouted up additional furniture, tablets, pens, and binders. Others strung up ropes to keep onlookers off the runways.

By late afternoon, some semblance of a command center was emerging from the chaos. Women were serving sandwiches and beverages. Rather than responding to incoming emergency calls willy-nilly, pilots were checking the official logs to see if a response might already be in progress. Instead of asking people at random where a particular volunteer might be at a particular time, anyone could check the logs for himself to see who was in the air. The ham operators set up a radio at the funeral home down the road so a hearse could be dispatched to meet any helicopter that was coming in with a corpse. Then Big Jim Tribble from the Office of Civil Defense returned from Amherst with six badly needed two-way radios and the news that food, water, fuel, and other provisions were on their way.

As dusk forced the aircrews to stop flying for the day, Cliff led his self-selected team of local volunteers to the old courthouse. With Jim Tribble's input, they drafted out a command structure that put Cliff at the top. No more buzzing around in aircraft for Cliff; he'd need to stay on the ground and coordinate all aspects of the response operations. One deputy was assigned to traffic duties, including allocating bumper stickers to restrict landing-strip access to authorized vehicles. Another volunteer was put in charge of coordinating rescues. A third would oversee the record keeping. The new school superintendent, Henry Conner, was to set up a relief center at the high school and supervise the distribution of food and clothing and the use of the school for temporary shelter. Doc Gamble would head up the medical team, whose duties included dispensing vaccines. Another volunteer would handle the task of body identification. So everyone would not be distracted by questions from pesky reporters, they designated a public information officer. And as fire chief they appointed the fire chief—one of the few county officials who was available to fulfill his regular duties.

That command structure was born out of need and designed out of expediency. Only those volunteers who were present that evening were given official responsibilities. Three of the four county supervisors were not included. And Sheriff Bill Whitehead was not included.

As the meeting closed, one of the locals commented that the ABC store (the state-run liquor retailer) was doing a brisker business than seemed to be in the community's best interests under the circumstances. Cliff started to promise that he'd check into the matter when he was interrupted by Jim Tribble. Big Jim suggested that they contact the ABC board immediately and just tell them to close the store. With that, they got a ham operator to raise another ham outside the area who in turn had a telephone patch to Lynchburg. They reached one of the ABC administrators at home and told him that the Nelson County Board of Supervisors wanted their local liquor store closed for an indefinite period. The next morning, a member of the ABC board radioed back and confirmed that the deed had been done.

Cliff was astonished. He'd never experienced anything in government that worked so efficiently.

Sheriff Bill Whitehead began Day 2 setting up the emergency center for the isolated Massies Mill District; when the Marine helicopter arrived, he joined in the aerial search and rescue efforts. He couldn't help but notice that the young pilot seemed edgy. When they landed briefly in Lovingston, Whitehead took him aside to give him a little pep talk.

The pilot broke eye contact and gazed off in the distance. He never thought he'd be in the position of having to apologize for the actions of his commanding officer, he said. This was embarrassing, but he and his crew had orders to return to Quantico the following day. They wouldn't be back.

Whitehead patted his shoulder and said nothing. Nearby, Cliff Wood was supervising the transfer of two bodies into a hearse. Three or four people were shouting for his attention from different directions. Whitehead jogged over. "I need another helicopter, starting tomorrow," he said. He thought he saw Cliff nod in agreement.

In the commotion, however, Cliff thought that the sheriff was asking for a total of two helicopters. As Whitehead reboarded the Marines chopper, Cliff mentally relegated the request to a low priority.

On Day 3, no helicopter showed up for the sheriff, even though several of them passed overhead. That afternoon, one landed near Massies Mill only to drop off a ham radio operator with his equipment and a sheet of paper on which the new command structure had been penciled. The sheriff's name wasn't on the list.

Whitehead was stunned that he'd been left out of the loop. His pride was wounded further when he heard that several high school boys were flying around as guides in army choppers. He continued about his tasks, coordinating the growing groups of volunteers searching the streambeds, trying to get emergency repairs made to the road washouts, and assuring that homeless victims had temporary shelter, all the while seething inside. In no emergency in the past had he ever run to the board of supervisors to get permission to discharge his official responsibilities, and regardless of the scope of the present crisis, he damn well wasn't going to do so now.

Then the ham delivered a radio message that had just arrived from headquarters. Cliff wanted to see Whitehead in Lovingston, as soon as possible.

It took Whitehead several hours to make the trip that normally took but fifteen minutes or so, but thanks to two bulldozers owned by a local excavation contractor, he got past the debris pile at the confluence of Hat Creek and the Tye River and continued on to the county seat. Unbeknownst to the sheriff, Cliff assumed Whitehead still had access to a helicopter and would simply fly there.

When Whitehead arrived at the command center, the Chesapeake and Potomac Telephone Company had just finished stringing in a few temporary long-distance lines. A pay phone hung on a stop sign, its coin box disabled, while three phones sat on a table beside the airstrip and another three were in the command center's tent. Calls began arriving nonstop. One of the first calls going out was that of the local undertaker ordering more caskets.

By this time, no more injured were being brought in. Only corpses were arriving, many of them grotesquely mangled. In the Vietnam War, there were typically six wounded for every fatality, but that was not the case here. Most of the flood victims had either survived with little or no bodily harm, or else they had drowned or were fatally crushed by the debris.

Still fuming, Whitehead sat across from Cliff at a folding table, expecting a remorseful apology for the shabby treatment and an affirmation of his status. Cliff began to ask a few questions but was interrupted repeatedly by people darting in with issues that couldn't wait. Shouts came from the phone desk: hey Cliff this, hey Cliff that. In what was the final straw for the sheriff, one of the volunteers sped into the tent on a bicycle, shouting and dismounting before he'd even stopped, and the bicycle got away and slammed into White-head's back. He stood up, all six feet seven of him, and glared at Cliff and the cyclist. If everyone else's business was more important, they could damn well get along without him. He marched back to his mud-splattered patrol car, started the engine, and drove off. Just as he left, Governor Godwin landed on the strip.

By now, enough makeshift road repairs had been completed to get jeeps and pickups into the Massies Mill District from Amherst County. Although the going wasn't easy, by late on Day 3, food, water, clothing, fuel, flashlights, batteries, and other provisions were arriving. Whitehead set up guidelines for determining need. Along with that aid, however, came the challenge of getting it to the folks who were still isolated. Meanwhile, they still had no means of searching for bodies along inaccessible streams. Whitehead swallowed his pride, radioed the command center, explained his predicament, and reminded the folks there that he still needed to be allocated his own helicopter.

News reporters began to arrive from the south and the west. Photographs of the devastation in Massies Mill made national news while other parts of the county that lay just as wrecked were ignored. The competence of the newspeople was highly variable, and many did not leave a good impression on the locals after poking microphones in the faces of victims who didn't want to talk and writing inaccurate stories of mountain folks living in tar paper–covered shacks. In contrast to the relief volunteers from the outside—most of whom brought their own food and provisions with them—many of the reporters fed themselves from the tables set up for the victims. Whitehead would later remark that one of his oversights was in failing to immediately set up roadblocks to keep out the reporters and curiosity seekers. A few days later, he remedied that situation.

At Lovingston, Cliff met Governor Godwin and several of his aides at the landing strip, and the combined group drove five miles north to the landslides at Woods Mill. An official photographer snapped a series of pictures of the governor brooding over the devastation.

There is always a degree of public cynicism—sometimes justified—about visits to disaster sites by the likes of governors and presidents. Everyone is aware that such inspections are obligatory performances; woe to any governor who might be insensitive enough to ignore a natural catastrophe within his or her state's borders. Such on-site inspections, however, do serve a function beyond the ceremonial. It takes a governor to declare a disaster before the president can do so, and that declaration in turn is a prerequisite to the flow of any federal aid.

His entourage in tow, Governor Godwin strode to a concrete walk that terminated at a vertical drop. "Was there a house here?" he asked. Cliff responded that, yes, four days ago a family had lived—he pointed—at a spot that was now but vacant air a few dozen feet beyond the ledge.

The governor contemplated the debris piles and the locals scaling

A landslide just north of Lovingston, Virginia. *(Courtesy M. E. "Ed" Tinsley.)*

them to pick out small items. He walked over to exchange a few words and then returned shaking his head. No, they weren't finding much of anything salvageable.

They drove back to the command center, Godwin directing his driver to stop each time he saw a local near the road. He repeatedly offered words of sympathy and ended each conversation with words of encouragement. At one point, the car stopped on a downed power line that two repairmen were about to raise. One of the workers shouted, "Get the hell off that damned wire!"

Godwin stepped out and introduced himself. The men stammered apologies. "Whatever you boys need," the governor said, "just tell Cliff here, and he'll get the word to me."

After his initial visit, Godwin returned on Days 4 and 5 and then at least once each week for the next month. On Day 4, heavy equipment showed up on the opposite side of the damaged bridge on the Rockbridge River, and a few days after that the crews had one lane of Highway 29 open to points north.

In his visits and by phone, the governor repeatedly offered to send in the Virginia National Guard. Cliff steadfastly declined, insisting that they weren't needed. He had dealt with a lot of unknowns so far, and although things hadn't necessarily gone as smoothly as he'd have liked, at least the local authorities seemed to be getting matters under control. Bring in the Guard, however, and who knows what might happen?

Years later, Cliff would candidly explain that 1969 was a time of student rioting on college campuses, antiwar protests in Washington, and race riots in cities across the nation. Wherever you saw the Guard, you saw big trouble. The present catastrophe was a local, not a state or national, disaster, and that's the way Cliff wanted to keep it. As far as having experienced leaders at his disposal, many of Cliff's local volunteers were veterans of World War II or the Korean War, and they understood about decision making and coordination. Cliff was confident that those old-timers knew what they were doing a lot better than would a bunch of National Guard enlistees arriving from the outside.

On Day 4, someone brought Cliff a bundle of newspaper clippings. To his chagrin, several of those articles portrayed the Nelson County locals as a bunch of hicks. From that point on, he refused to have any interaction with the reporters. If one tried to collar him, Cliff sent him off to talk to the command center's public information officer, who would invariably be off doing something else.

Meanwhile, Cliff toughened up on air traffic. With so many relief provisions being airlifted in, and all of the rescue teams being transported to and from remote areas, it was time to put a halt to sightseers whose flights were contributing to the congestion. He put out the word: If an aircraft should show up that has no relationship to the ongoing relief activities, it was not to be given permission to land no matter who was on it—even if it was Walter Cronkite. The next day, what started out as someone's joke about Cliff's comment developed into a rumor that Cronkite had indeed been denied landing privileges at Nelson International.

As of Day 5, Bill Whitehead was still waiting to be allocated one of the helicopters that were buzzing around the command center's airstrip in Lovingston. It didn't help his level of frustration that he was getting reports of choppers landing at various spots around Massies Mill, some self-styled outside expert getting out and puffing around with nothing in particular to do and generally annoying everyone, then taking off again. When one of the county's preachers landed near the sheriff's parked car, Whitehead asked him if, at last, this was the helicopter he'd requested. No, said the minister, this was *his* helicopter. Whitehead's reaction was something less than diplomatic, and the preacher carried the news back to Lovingston that the sheriff wasn't very happy about the situation.

On Day 6, Whitehead returned again to the command center tent, where he got a chilly welcome. In a booming no-nonsense voice, he reminded Cliff that he needed a helicopter. Cliff shouted back that Whitehead had already gotten a Marine chopper five days earlier and that he had been ignoring the chain of command and running off half-cocked doing things on his own. Dozens of people within earshot fell silent to watch the altercation.

Austin Embrey, clerk of the circuit court, ushered the two men out behind the tent, where the conversation might have some chance of proceeding without an audience. Knowing better than to leave the two strong-willed men by themselves with their tempers flaring, he stuck around to "take notes."

The shouting match escalated, Bill Whitehead and Cliff Wood both expressing things in the heat of the moment that both would later claim they had forgotten. Neither had gotten a full night's sleep in five days, and both were exhausted to the point of collapse. And then it hit them both, almost simultaneously, that the whole problem had arisen out of a miscommunication. Cliff had jumped to the erroneous but understandable conclusion, after seeing the sheriff in a helicopter on Day 1 and Day 2, that he had no need for an extra one. The sheriff, in his preoccupation with the crisis, had neglected to explain the details of the helicopter situation that would have prevented the misunderstanding. Cliff, in any case, had been so inundated with his own issues at the time that he may not have been attentive anyway. And so on.

They looked into each other's eyes. Whitehead found himself choking back a tear, and Cliff was on the verge of crying as well. They shook hands, agreeing that they each bore their respective portions of the responsibility for the conflict, both having jumped to flawed conclusions about the other's motives. They were barely started on the long road to recovery from the disaster, and never again would they allow a misunderstanding like that to undermine their ability to cooperate in the serious work that lay ahead.

CHAPTER 15

OUTSIDERS

Initiated more than a century earlier by the condescending meddling of the Northern abolitionists, then exacerbated by the hated carpet-baggers, a mistrust of outsiders had been passed down from generation to generation throughout the rural South. A shadow of suspicion is evident even in the way strangers are greeted: "Y'all aren't from around here, are you?" Such deep-seated wariness does not instantaneously evaporate in the wake of a disaster, and it was bound to fuel apprehensions about the motives of even the most well-intentioned of outside relief workers.

As the scope of the disaster became apparent, Camille's victims experienced a dawning recognition that their needs overwhelmed the capabilities of their close-knit communities. They knew they couldn't possibly fix everything themselves, yet a part of them felt humiliated about accepting aid from outsiders. The tensions between pride and gratitude, xenophobia and hospitality, and self-reliance and authentic needs muddied many of the interactions between natives and the relief workers—especially when the relief emanated from the feds.

The Salvation Army, the Mennonite Disaster Service, the Red Cross, and a long list of other charitable organizations and churches quickly dispatched relief teams. Closely following were representatives of an alphabet soup of federal departments and state agencies. Although each bureaucratic unit had its own mission and recipient eligibility criteria, confusion arose from the numerous overlaps, conflicts, and oversights in the regulations. Many officials, though well trained in their specific areas, were largely ignorant of what other agencies could and couldn't do. As a result, numerous victims got a runaround from one bureaucracy to another. Compounding the mayhem was the influx of insurance adjustors, some prepared to make quick and minimal cash-in-hand settlements in return for the waiver of further claims by the still shell-shocked victims, most of whom didn't understand what they were entitled to.

It was far from a simple matter for survivors to find the temporary field offices of the various agencies; the victims were without power or phone service, no mail was being delivered, roads were impassable, and even when they thought they knew where to go, simply getting there posed challenges. Signs posted on power poles didn't always convey the intended message. Many victims assumed, for example, that an agency with the words "Small Business" in its title couldn't possibly have anything to do with loans for private home repairs, when in fact that was a big part of what the Small Business Administration (SBA) was there to do.

Similarly, although the Army Corps of Engineers was restricted to restoring those parts of the physical infrastructure whose condition threatened public safety, it was possible for some farmers to legitimately claim that clearing or rechanneling streams on their private property met the Army Corps of Engineers requirements. Some of the representatives from Housing and Urban Development (HUD) were not aware of all they were empowered to do and gave out erroneous information. According to a provision passed by Congress just a month before the disaster, for instance, recipients of temporary housing were not to be charged rent for the first three months. Unfortunately, that policy was still another few months from appearing in the next year's HUD regulation manuals, and many of the field-workers didn't even know about it.

The disaster did, however, bring a new dimension to the military's relationship with local civilians in Mississippi. Cynics might say that the armed services were only seizing the opportunity to improve their images in the face of the eroding public support for the Vietnam War. Regardless of the motives of the top brass, however, the boots on the ground proved invaluable to the stricken communities. Two planeloads of Seabees returning from Vietnam arrived in Gulfport a day after the storm and were immediately dispatched to the devastated neighborhoods to operate front-end loaders, dumptrucks, and chainsaws. At Keesler Air Force Base, hundreds of young airmen were sent into the civilian sections of Biloxi to help with the cleanup. The army sent several squadrons of helicopters to the devastated regions and put them and their crews at the disposal of the civilian authorities. And the Coast Guard got to work clearing the navigation channels and replacing the buoys and other navigational aids.

Wandering around in the rubble were thousands of confused pets, unable to locate their homes or their families. Providing food and water for all of them was impossible. Following the declaration of martial law on August 19, military personnel in Mississippi were ordered to destroy any lost animals they encountered. Young soldiers stifled tears as they dispatched sad-eyed dogs to canine heaven.

Fire ants, which can survive for several days underwater, emerged everywhere in the debris. Young military men who had grown up elsewhere and didn't know any better found their hands, arms, and ankles covered with festering stings. Allied Chemical Company donated fifty tons of Mirex, an insecticide now banned by the Environmental Protection Agency, and as soon as the Gulfport municipal airport had an open runway, crop dusters began spraying the toxic chemical over seventy-five thousand acres along the coast.

The goodwill generated by the military, however, was largely undermined by other agents of the federal government. When you're averse to asking for help to begin with, you become particularly sensitive to anything that smacks of arrogance and incompetence.

During the postdisaster Senate hearings, Senator Bob Dole asked Mississippi's regional director of the SBA whether his staff had been able to offer relevant guidance to all of the prospective applicants. The director acknowledged that this had been a problem and offered as an excuse, "Really, there are so many programs available, we get confused ourselves."

Dole's question did not arise from a vacuum; his office had gotten numerous complaints about the SBA's performance. HUD officials also took a beating. A number of low-income victims had apparently been advised that they needed to have their own lot and pay for their own utility connections if they wanted HUD to rent them a mobile home for a year. In fact, the official policy was that, *if* they owned a private lot, then HUD expected them to provide their own utility hookups; otherwise, the rental unit would be set up on municipal property and HUD would connect the utilities.

It was not just the prospective beneficiaries, however, who experienced frustrations with the regulations and their interpretations. Several HUD administrators complained that they were forced to go through a complicated federal procurement procedure to buy mobile homes directly from the manufacturers, when in fact hundreds of new mobile homes were already available on dealers' lots within a few hundred miles of the disaster zones. Those dealers would have happily supplied the manpower and expertise to deliver and set up those housing units. If HUD had tapped this readily available inventory, many of the homeless could have set up housekeeping within a few weeks instead of waiting for several months.

Among the other witnesses at the Senate hearings was Robert C. Goad, commonwealth attorney for Nelson County, accompanied by Cliff Wood and three other county officials. Goad was far from complimentary to the federal bureaucracy, stating for the record, "The Office of Emergency Preparedness is not prepared for an emergency!"

Goad went on to describe the red tape that had hampered the ability of local and state officials. One egregious example: Office of Emergency Preparedness (OEP) Circular No. 4000.4A, supposedly an assistance manual, contained at least 169 pages of rules and regulations governing relief activities. The nature of an emergency, however, requires that decisions be made quickly and actions be taken at once to save lives and property, without requiring community leaders to first take the time to digest 169 pages of tiresome regulatory prose.

Further, those regulations made it clear that the OEP was oriented to urban disasters, not emergencies in rural regions. For instance, engineers quickly made detailed assessments of the damages to Nelson County's water and sewer systems and recommended to the OEP that they authorize payment for the necessary repair work. The OEP rejected this request by the engineers they themselves had hired. Authorizations could be made for repairs only to public systems, and Lovingston was served by a privately owned water system run by the pharmacist and an undertaker. Both of those men, of course, were busy with other matters after the disaster. The glitch was this: although classified as a public utility under Virginia law, the water system did not qualify as a public utility under federal regulations.

The Nelson County contingent pointed out that it was not unusual for rural water systems to be privately owned. And when a disaster strikes, rural people need water and sanitation and other public services just as much as city folk do. Regulators needed to quit assuming that emergencies arise only in urban areas.

A repeated theme throughout the testimony was that the locals needed resources and coordination from Washington, *not* rules and authority. Natives knew best what it would take to solve their problems, and for the federal government to behave otherwise was inefficient, condescending, and presumptuous.

As a result of those Senate hearings, it became obvious that future responses to natural disasters needed to be coordinated, although not necessarily managed, by a single agency. Thus, the precursor to the Federal Emergency Management Agency (FEMA) was born out of the Camille disasters. Names of organizations can be telling, however, and although the locals kept emphasizing they didn't want the federal government to *manage* disaster response and recovery, "management" became part of the new agency's name. Interestingly, however, we've recently come full circle, because FEMA is now under the province of the Department of Homeland Security—making it once again interconnected with civil defense.

While most of the federal agencies focused on restoring the infrastructure and otherwise laying the groundwork for long-term recovery, the private and semiprivate relief organizations addressed emergency needs: water, food, clothing, sanitation, and public health. Although funded through private donations, the Red Cross had been playing a semipublic role in disaster relief since 1905, when—in an arrangement that remains essentially intact to the present day—Congress chartered that organization to coordinate emergency relief in federally designated disaster zones.

After Camille, however, numerous survivors testified that they were grossly displeased by the performance of the Red Cross. According to one widely circulated account, a Salvation Army truck was stopped as it entered Gulfport, the Red Cross transferred its load of food to one of its own vehicles, and then it trucked those foodstuffs off to a warehouse in Texas. The Red Cross did not deny the incident. The rationale was that too much free food distribution in the disaster zone would be economically injurious to the local food retailers. Instead of direct food handouts, victims along the Gulf Coast were given Red Cross food vouchers that they could redeem at local grocery stores.

Unfortunately, the Red Cross staff hadn't taken the time to confirm that economic theory by conducting a census of the surviving supermarkets. In the worst-hit communities of Long Beach and Pass Christian, no grocery stores had survived. A victim in that area testified that she had hiked more than two miles across the damaged bridge to the nearest food market, where she rudely discovered that her Red Cross grocery vouchers could not be honored because they were good only in Harrison County.

When Governor Godwin made one of his repeat visits to Nelson County, a driver jumped into a Red Cross truck, positioned it where its bright red logo would stand out on the news footage, and then began forcing sandwiches on passersby whether they wanted them or not. When the camera crew packed up, the sandwich charade ended. The implication to many of the onlookers was that the Red Cross was interested mainly in the publicity that would bring in more private donations and really didn't care enough about the stricken natives to reach out and assure that the aid was reaching those who needed it most. Although the Red Cross would later admit to the validity of some such criticisms, and would make corresponding adjustments in its operating practices, for many Camille survivors the reputation of that organization was badly tarnished.

In contrast, natives gave the highest possible marks to the Salvation Army and the Mennonite Disaster Service, both of which went into the hardest-hit areas to take on the most difficult jobs. In Virginia, the recovery of bodies was a particularly challenging task, scores of the dead having been washed many miles downstream where they were buried in mud or tangled up in massive stacks of debris. For four weeks, Mennonite men went out daily with the search teams, tramping along the mud-covered riverbanks with shovels and saws. Each day, the corpses they found were in further stages of decomposition. Even after a body was discovered, sometimes it took an entire afternoon to retrieve it, and it was only the Mennonites who seemed able to do that kind of awful work relentlessly, including transferring corpses into body bags and carrying them to the nearest site where a helicopter could land. As the Mennonite men were engaged in this gruesome labor, some of their wives were busy in the villages, shoveling mud out of surviving homes, cleaning sidewalks, and salvaging furniture and other useable or repairable items.

The Mennonites worked diligently and quietly, never uttering a word of complaint and never proselytizing their religion. They sought no publicity, not even for their church, and it didn't seem to bother them that their work went largely unheralded in the press. When the search operations were terminated in late September, they returned to their farms in places like western Pennsylvania, Ohio, and southern Ontario, promising to return after the fall harvest.

And return they did, bringing truckloads of tools and building supplies with them. (Unlike the Amish, who shun motor vehicles and power tools, the Mennonites are very practical about the best ways to get things done.) Over the next several months, they repaired or rebuilt dozens of Nelson County houses at no cost to their owners. One of the homes they worked on was that of Warren and Carl Raines.

The immediate unspoken goal after any disaster is to restore normality—to get things back the way they were and as quickly as possible. The fact that this is always an impossibility (no society can ever turn back the clock, particularly after a catastrophe) does not stop communities from trying. After Camille, the objective of restoring the status quo was sometimes carried to ridiculous extremes, as when the Red Cross in Mississippi provided aid differentially according to the victims' social status: if they had an outhouse before, they got a new outhouse after; if they shopped at used clothing stores before, they got used clothing after; but if they bought everything new before, then the Red Cross gave them vouchers for new clothes and furniture. This policy, of course, perpetuated existing class stratifications.

One white woman who entered a clothing distribution center with two black friends (not all Mississippians were racists) found herself pulled aside by a white volunteer who led her to a rear room where the "good clothes" had been set aside. In another instance, a black woman was given food vouchers that amounted to little more than half of what had been given to a white woman with the same size family. Such disparities in aid distribution were consistent with the contemporary segregationist culture in Mississippi.

Blacks were hard-hit in Mississippi, not because they lived in any of the upscale homes near the beach, or even in the second-tier middle-class homes, but because Camille's monstrous flood had surged up the inlets, bays, and bayous, circling in behind the coastal towns. Most of the black community on the backside of Biloxi, for instance, was inundated by four to six feet of floodwater.

Dr. Gilbert Mason, the man who had initiated the court battle to desegregate the beaches and who lived in the community he served in his medical practice, was hit right along with the rest. Although his home stood nearly eleven feet above sea level, it was flooded to a depth of four and a half feet while he and his family huddled in their attic. With their clothing ruined, their two cars demolished, and all of the local banks and other businesses closed, they were just as destitute as the rest of the disaster victims.

Dr. Mason began treating the injured as best he could with the minimal available supplies, looking a bit bizarre in his flip-flops and a bright blue jumpsuit. When the Red Cross set up a clothing distribution center in Biloxi a few days later, Mason walked there to get himself a pair of shoes. To his chagrin, he was summarily denied on the basis that, if he was indeed a doctor, then he didn't need any help. Later, he would write:

> Red Cross workers in Biloxi developed a reputation for need-lessly humiliating those seeking relief, especially black folks, whom they often seemed to make beg, cajole, or grovel for assistance.

It's unlikely that Dr. Mason was just being a grump; at the time he penned the above assessment, he simultaneously gave high marks to the efforts of the Seabees, the airmen from Keesler, the Salvation Army volunteers, and even the Biloxi Housing Authority, with whom he had butted heads on numerous prior occasions for its discriminatory practices.

A more cooperative racial climate was certainly possible, as exemplified by the relief programs in Virginia. Governor Linwood Holton, who succeeded Mills Godwin in January, assigned a young African American aide, William B. Robertson, to ferret out any evidence that relief programs—private or governmental—had treated black and white flood victims differently. Although he subsequently documented a variety of shortcomings and disparities in the distribution of aid, Robertson testified at the Senate hearings that he was unable to find any evidence that the inconsistencies ran along racial lines. Nelson County and the other affected counties in Virginia may not have been totally color-blind, but at least race had played no major part in who did or did not receive postdisaster assistance.

Meanwhile, another group of outsiders had a major impact on the recovery: the U.S. Department of Justice. Camille, by a remarkable coincidence, had chosen to flatten three of the very places that the feds had been strong-arming, fifteen years after *Brown v. Board of Education,* to desegregate their schools. Nelson County had integrated the primary grades the year before, and the high school was to be integrated when school began in September. Blacks and whites alike were nervous about the transition to a fully integrated school system, and leaders from both races had been working together for the past two years to assure that the planned full desegregation in 1969 took place as smoothly as possible. In a respectful tone, one of the white leaders characterized the black leaders as "no-nonsense kinds of guys." School integration was going to happen regardless of whether all of the natives supported it, and for the sake of everyone, it needed to work. It did.

The white elementary school and the black schools in Pass Christian were too severely damaged to hold classes, which necessitated the integration of all grades that fall. Temporary trailers were set up on the grounds of the formerly white high school to accommodate every student of both races. Everybody knew there was no choice about it, especially now that Camille had made sure there were no black schools available. Perhaps the students' common experience of the shared disaster had something to do with it, but when black and white teenagers began sharing classes for the first time ever, taught by both black and white teachers, there were no disruptions worth noting.

Plaquemines Parish, however, was another story. Before Camille struck, the public education system was already in chaos as a result of Leander Perez's rabid resistance to integration. Even the hurricane of the century would not expedite the desegregation process there, as it did in Mississippi and Virginia. It would be another two years before the schools began to recover from the double whammy of the Perez legacy and Hurricane Camille.

Not counting the Mississippi evacuees who had yet to return, Camille left at least fifty-five hundred penniless and homeless victims wandering around the stricken Gulf Coast with few places to seek shelter. In one of the few executive decisions he would make while in office, Governor John Bell Williams ordered the full postdisaster evacuation of Pass Christian and Long Beach. As soon as the roads were clear enough, a parade of state-rented buses rolled in.

Under martial law and the governor's order, the newly rendered homeless had no choice but to leave the coast. The majority of those falling under the order happened to be low-income, and now-jobless, blacks. About fifteen hundred of these refugees were bused up to Jackson on racially segregated buses, where they were joined by a smattering of others who had managed to find their own transportation. Whites were put up at the Robert E. Lee Hotel, but blacks were sent to Jackson State University, a black institution. A few blacks, however—George Watson, the principal of the Randolph schools in Pass Christian, among them—vehemently insisted that they also be put up at the hotel. The hotel management responded by allocating the top floor for black folk.

After a week, those who had not yet found alternative housing in Jackson were transferred—again on segregated buses—to Camp Shelby, near Hattiesburg, which meanwhile had been accommodating most of the other four thousand postdisaster evacuees. Because that base was federal property, the races were automatically integrated once the refugees set foot inside the gate.

Camp Shelby's main use was to train National Guard troops during the summer, and it hadn't been designed to provide many creature comforts. It had never been intended to accommodate families, and the barracks were anything but private. The camp commander saw little choice but to separate the refugees into different barracks by gender. Some of the displaced blacks, including George Watson, took offense that their families would be broken apart. Mr. Watson, whose personal car had somehow survived, drove his family back to their damaged home, shoveled out the mud from one small room, and took refuge there without any utilities or other creature comforts and without even any windows to protect them from the mosquitoes.

When Dr. Mason heard about the segregated evacuation buses, he was furious. Just days earlier, some of these same blacks and whites had pulled each other out of the water or from trees, and now Governor Williams didn't think they should even share the same bus?

This was the same John Bell Williams who, as a U.S. congressman a decade earlier, had delivered a diatribe in the U.S. House of Representatives in which he cited soaring black crime rates, claimed that the number of serious offenses by blacks was highest in racially mixed neighborhoods, and concluded that segregation was the only answer. This was also the same man who later supported the racists who fought Dr. Mason's attempts to integrate the Gulf Coast beaches.

If it seemed the disenfranchisement of blacks couldn't get worse, it almost did. To the fanfare of scores of political supporters, Governor Williams appointed a special Governor's Emergency Council to oversee the reconstruction of the stricken counties. All of the appointees were white males, and they held their meetings up in Jackson, not down on the affected coast. Dr. Mason complained immediately, not just to the governor's office but to Washington. Recovery planning, he argued, should not ignore any significant segment of the stricken community. The governor responded that his council members were all well aware of the needs of the black folks, and he didn't understand what all the fuss was about.

At the beginning of January 1970, Mason unexpectedly received word from the governor's office that the membership of the Governor's Emergency Council had been expanded to include one woman and three black men. Mason was one of the additions. Curiously, his appointment predated by a mere few days the commencement of hearings in Biloxi of the U.S. Senate special subcommittee on disaster relief relating to Camille, hearings at which Dr. Mason had already been scheduled to testify.

In 1969, meteorologists from the National Hurricane Center seldom conducted firsthand inspections of the sites of hurricane landfalls; if and when they did so, such excursions were on their own time and had nothing to do with their official duties. As for the NHC's directors, prior to Camille none of them had *ever* visited the site of a hurricane disaster in the immediate aftermath of the event.

Camille, however, had broken all records for meteorological severity. Dr. Bob Simpson wasn't the kind of bureaucrat who might simply relegate this event to the archives without first tramping through the debris on his own feet, examining the devastation with his own eyes, and hearing a sample of survivors' stories with his own ears. Simpson flew to Gulfport but, unlike President Nixon, didn't confine his visit to the airport. In Pass Christian, Long Beach, and Bay St. Louis, he hiked through miles of wreckage—taking notes on the devastated buildings, the beached shipping, and the wrecked infrastructure of roads and utilities. Thinking, thinking, thinking.

The meteorologists at the NHC, Simpson reflected, couldn't possibly be familiar with the particulars of every community a hurricane might strike. Nor, given the state of scientific knowledge about hurricane behavior, could the NHC predict a landfall with any significant degree of advance notice. Given this quagmire of uncertainty, he asked the local authorities what specific information the NHC could possibly provide that would assist them in making future evacuation decisions.

In response to this repeated question, Simpson found himself hearing a recurring theme from a variety of people: There are hurricanes, but then there are *hurricanes*. Folks need to know something about the storm's strength—its capacity to destroy.

Although the NHC advisories and bulletins always included the current wind speed information, it became clear to Simpson that this numerical data didn't satisfy the real needs of the local decision makers. What was needed was some simple but meaningful connection between a hurricane's wind speed and its threat to human life and property. Simpson returned to his Florida office and mulled over the issue.

Although Simpson was an expert meteorologist, he knew no more than most other educated laymen about the structural integrity of buildings in high winds and flooding. Meanwhile, Dr. Herbert Saffir, a civil engineer, had been thinking about a hurricane intensity scale that was based on the wind loading on structures but that also needed to be linked to what was known about tropical meteorology. Saffir had worked in southern Florida since 1947 and had studied the destructive effects of dozens of hurricanes from a civil and structural engineering perspective. An outspoken advocate of stricter building codes, he had started his own consulting firm, took on a variety of municipalities as clients, and conducted a study for the United Nations on hurricane hazards to low-cost housing.

Saffir's engineers demonstrated that sturdier buildings could be constructed—even homes for people of modest means—without bankrupting anyone. It was his continual lament that, even when communities did enact strict building codes, enforcement was often lax. Many folks got away with it because sometimes even shoddy construction would survive a hurricane. Not a *bad* hurricane, of course, but then in 1969 there was still no scale that distinguished hurricanes according to any kind of an intensity scale. As a consequence, the survival of a poorly constructed building in a minimal hurricane was often erroneously taken as evidence that the construction was of a sufficient standard to withstand hurricanes in general.

Herbert Saffir drafted a five-category damage potential scale and sent it to the NHC with the hope that Bob Simpson would take a look at it and give him some feedback. His letter serendipitously arrived right after Simpson's inspection of Camille's devastation along the Gulf Coast. Although Saffir's focus was different than Simpson's—Saffir was thinking about building codes and survivability while Simpson was thinking about evacuation warnings—their ideas meshed remarkably well. A single "damage potential scale" could address both objectives.

Alphabetically, Simpson's name went second. Starting with Herbert Saffir's correlations of wind speeds with types of easily observable structural damage, Simpson drew on Chester Jelesnianski's seminal work on storm surge prediction and created the five-category scale everyone is familiar with today. Simpson explicitly acknowledged the importance of Jelesnianski's work in creating the scale, and perhaps if that Polish American meteorologist's name had been shorter and more pronounceable, it would have entered the lexicon as part of a "SSJ Scale."

The Saffir-Simpson Scale made its debut in the 1971 hurricane season, less than two years after Camille. It was a simple, easily understood method of alerting emergency planners and the public at large of the *potential* threats from an oncoming storm. Herbert Saffir and Bob Simpson's numerical criteria for hurricane categories remain today essentially as they were originally developed in 1970 (see appendix).

Simpson and his staff mulled over a variety of issues as they injected their meteorological descriptions into Saffir's preliminary damage scale. Should they, for instance, even bother to include the barometric pressure when it is the wind that causes the damage (either directly or by virtue of whipping up the sea)? Wouldn't the wind speed information alone be enough to establish the category? Although there is a correlation between the central pressure and the sustained wind speed in a hurricane, that relationship is complicated somewhat by the physical size of the storm. Tightly wound hurricanes such as Camille tend to have higher wind speeds than bigger hurricanes with the same barometric pressure.

Ultimately, Simpson and his staff decided that the barometric pressure should stay. Although wind speed would remain the main criterion, winds are difficult to measure and anemometers are notorious for disintegrating during hurricanes. Barometric pressure, on the other hand, is easily measured indoors with a static instrument. Even in the absence of direct wind speed measurements, it would still be important to be able to establish the category of a hurricane (and therefore its damage potential). If a storm's central pressure was 28.10 inches, for instance, even in the absence of a wind speed reading one could be pretty sure that this was a Category 3 hurricane with its attendant potential for inflicting Category 3 damage.

Using this set of criteria, the staff at the NHC went back through its archives of historical storms and came up with just one Category 5 storm prior to Camille. Based on its barometric pressure, that was the so-called Labor Day Hurricane, which devastated the Florida Keys in 1935 and for which there is no wind speed data whatsoever. Hurricane Andrew in 1992 did not meet the Category 5 barometric pressure criteria, and its wind speed was on the cusp of a 4 and a 5, but because of its catastrophic damage, ten years later in 2002 it was recategorized from a Category 4 to a Category 5.

Camille, the hurricane that gave rise to the Saffir-Simpson Scale, remains as of this writing the only hurricane that has ever met all of the Category 5 criteria at the time of its U.S. landfall: sustained wind speed in excess of 155 miles per hour (officially 172 miles per hour but possibly as much as 201 miles per hour with even higher gusts), barometric pressure lower than 27.17 inches (officially 26.84 inches), and a storm surge greater than 18 feet (officially 24.6 feet but probably closer to 28 feet).

Herbert Saffir was once asked why the scale stopped at Category 5. Might a superhurricane come along someday that would require extending the range to a Category 6 or even a 7?

Nope, Saffir responded; as a potential damage scale, 5 is the limit. Once you have Camille-like destruction, there is virtually nothing left standing, and the devastation can't possibly be any worse. Higher winds and surges would do nothing more than further stir the rubble.

Robert Simpson and his meteorologists agreed, although for slightly different technical reasons. There is a natural limit to how far the barometer can drop at sea level, and although that constraint depends somewhat on the latitude (it is slightly lower nearer the equator), Camille flirted with the bottom limit at the coordinates of the Mississippi coast. It is theoretically impossible for a hurricane to be much more intense than Camille, because the mechanics of atmospheric behavior simply will not permit it.

Perhaps at one level it is reassuring that Camille was the ultimate tropical cyclone and that such storms can't possibly be more severe. The flip side, however, is that the Category 5 designation describes not merely a possibility of such extreme storm destruction but the *reality* that tropical storms occasionally do get this violent. The warm waters of the Gulf and the South Atlantic are certain to whip up other Category 5 storms in the future—an average of two or three per century.

CHAPTER 16

A KNOTTY LEGACY

How many died? Any attempt to determine the *exact* death toll of a major disaster is fraught with difficulties. Some of the bodies are never found. Hapless travelers just passing through may succumb without loved ones tying their disappearance to the disaster. Further complicating matters is the cause of death. If someone dies in a traffic accident during an evacuation—before the storm even strikes—is that person a victim of the hurricane? How about a man electrocuted by his own emergency generator the day after? A child bitten by a displaced water moccasin? A woman dying from a ruptured appendix because she couldn't get to a hospital through the debris-clogged roads? To avoid such judgment calls, many authorities tally only those fatalities that can be directly attributed to the agents of destruction—in Camille's case, water and wind. As a result, deaths in natural disasters are often undercounted.

Camille's official mortality statistics emerged slowly and piecemeal. Six months after the disaster, the New Orleans District of the Army Corps of Engineers reported 262 confirmed deaths, but that figure did not include those still missing. At about the same time, the Virginia District reported 152 dead or missing in its jurisdiction, including 124 in Nelson County alone. The Biloxi *Sun Herald* quoted a figure of 172 deaths in Mississippi, which included 131 bodies recovered and 41 people missing, most likely washed out to sea. Combining what seem to be the most reliable figures results in an overall death toll of about 335. Other writers have arrived at a slightly higher or lower total, but regardless of such variations, Camille's death toll was remarkably low, considering the violence of the storm.

Although it might be supposed that compiling statistics on disaster fatalities accomplishes little beyond satisfying morbid curiosity, the death toll from a disaster informs us, at the very least, about the effectiveness of emergency planning and hazard mitigation. Clearly, for each person whose life was lost, mitigation was either nonexistent or ineffective. Thus, the number of deaths relative to the size of the at-risk population gives us a measure—albeit an imperfect one—of that population's state of preparedness for the disaster.

On the oil platforms, where every one of the thousands of workers was taken off in a timely manner, there were no fatalities at all. In Plaquemines Parish, where virtually everybody evacuated the lower peninsula, there were only seven deaths—fewer than one fatality per twenty-five hundred residents (not all of whom, of course, were equally at risk). Along the Mississippi Coast, where more than eighty-five thousand people evacuated and another fifty thousand or so took at least some steps to prepare themselves, fewer than two hundred died—a mortality ratio of less than one death per seven hundred natives. In Nelson County, Virginia, however, where no evacuation warning was possible, the death rate exceeded one fatality per *one hundred* county residents.

The demographic distribution of the deceased may also tell us something about possible ethnic, class, or age bias in a mass evacuation. Here, once again, the three regions devastated by Camille fared pretty well: in each place, the proportion of black fatalities (perhaps by happenstance) was no greater than the percentage of blacks in the regional population. In Virginia, blacks were actually underrepresented among the victims. Despite the long history of racial discrimination in Plaquemines Parish, Luke Petrovich saw to the evacuation of blacks and whites alike, even mixing them on the same buses and in the same shelters. In Mississippi, where racial tensions also ran strong and no provision had been made to evacuate thousands of poor blacks prior to Camille, local shelters were likewise integrated. The shameful segregation of the post-disaster evacuation buses in that state was ordered by the governor's office, not by the local officials in the disaster zone.

The obvious lesson is that lives indeed are saved when local officials and at-risk residents have time to prepare. Proper preparation, however, requires knowledge that cannot be generated using local resources. The National Weather Service, the National Hurricane

Center, and the Army Corps of Engineers (which monitors the river gauges) are essential national resources, worthy of taxpayer support, even by folks who otherwise claim to oppose big government.

Death statistics are one thing; the gruesome reality of recovering and identifying dead bodies is quite another. And in 1969, DNA testing was still not on the scientific horizon.

In Mississippi, three of the dead were never identified and were buried as "Faith, Hope, and Charity." In Nelson County, eight bodies remained unclaimed—despite Lovingston dentist George Criswell's having circulated their dental records to over ten thousand dentists nationwide. Those unidentified victims included three children, a boy about seventeen years old, a man in his late fifties, a woman in her mid- to late thirties, a woman in her late forties or early fifties, and a woman in her seventies. At least three of them had identical stomach contents, suggesting that they had shared their last meal together. Their descriptions did not match any of the thirty-two missing residents of Nelson County, and in a community that small, somebody would have known them had they resided there.

Could it be that nobody ever searched for those people? Or did family members search in another state, never dreaming that their loved ones could have been victims of Camille? Eventually the eight unclaimed Virginia bodies were cremated and the ashes sent to the state medical examiner's office in Richmond. Their ashes were never claimed.

As Camille drifted out to sea to die a quiet death of her own in the North Atlantic, she left in her wake three wretched swaths of dead and injured people, splintered homes and businesses, inundated farmland, livestock carcasses, impassable roads and bridges, power outages, and communications failures. In all three of the devastated regions, men and women who had been accustomed to living quiet daily lives found themselves catapulted into roles of major public responsibility. Even Igor Sikorsky, a Russian-born engineer who had never been to the American South, was transformed into an unsung hero as scores of his brainchildren—helicopters—saved hundreds of lives and delivered essential provisions that could arrive no other way.

Panic was rarely more than a short-term reaction, even among children. Although earlier publications about disasters were replete with stories of mass hysteria, looting, scapegoating, aggression, and other maladaptive or antisocial behaviors, the most typical human response to Camille was actually stunned silence as the enormity of the circumstances sank in. Then, relatively quickly, most people tended to regain their self-composure and began to assist each other. Such socially convergent behavior was the norm throughout all three stricken regions. Reduced to the level of their fundamental humanity by the crisis, Virginians, Louisianans, and Mississippians—black and white alike—were not all that different from one another.

In all three places, the initial search, rescue, and relief work was undertaken by the disaster victims themselves, long before outside aid began to arrive. Temporary leaders emerged, then, after their brief stints in the limelight, most of them quietly returned to their pre-disaster lives as best they could. Survivors in the three disaster zones mostly refrained from blaming the catastrophe on anyone—not even on the folks at the NHC or the Weather Service. Natives accepted the catastrophe as something that had simply happened, something that they needed to deal with, and which required their pulling together. In the *immediate* aftermath, class, race, and political distinctions evaporated. The common good transcended such petty differences.

It is a consistent feature of disasters that the threat comes from *outside* the community, and that victims quickly concur on the causes. People in the throes of a rapid-onset disaster don't argue much about the "facts"; everybody pretty much has the same admixture of information and ignorance. In the acute phase of a disaster, the immediate needs are easy to agree upon—so easy, in fact, that no discussion may be called for at all. This contrasts with more durable threats to the social order such as crime or joblessness, which tend to produce different sets of facts with conflicting interpretations. As a result, the initial phases of disaster response usually create less fuel for political conflict than do other types of social challenges. Later phases of recovery, of course, may be a different story.

In the news, a disaster may dominate the national scene for a few days to a few weeks. It will continue for a few additional weeks in the regional media, and then begin to dwindle except for the ritual of retrospection on each passing anniversary. For victims, however, the effects of the disaster may drag on for years. The disaster may even become the defining event in some people's lives.

Many news reporters, and even some academics, confuse the cohesive performance of social groups after a natural disaster with the strengths of their individual members. The public image of a man sharing his food and joining search parties when his own farm has been damaged, may not, however, be an accurate indication of what is going on inside him. In reality, group behavior in the throes of a crisis can be quite at odds with the internal states of minds of its constituents. Individual survivors can be emotionally devastated, torn apart by personal loss and sick with worry, yet function as amazingly resilient and productive members of their community during the acute response phase.

We are social animals, and programmed deep in the genetic makeup of most of us is a commitment to the survival of the society of which we are members. When that society has been critically damaged, restoring it to normality rises to a top priority. Disasters thus become occasions for renewing and strengthening bonds not just within the family, but also with the broader community. Those who misunderstand this intrinsic characteristic of human nature may erroneously reckon that certain individuals are coping just fine psychologically when actually they are not.

The unfortunate flip side of this issue is that, particularly in densely populated urban areas, some of the socially marginalized victims may not view themselves as part of the society they rub elbows with daily. When disaster strikes such a place—as Katrina did in 2005—pockets of lawlessness can indeed break out. This does not negate the aforementioned conclusions about social convergence; it merely means that the local and state authorities need to be aware of the specific characteristics of their communities, and plan accordingly.

After Camille, most survivors experienced immediate stress reactions: shock, fear, feelings of emptiness, confusion and disorientation, indecisiveness, inability to concentrate, memory loss, fatigue, insomnia, edginess, and feelings of abandonment. Most quickly recovered. For others, particularly those who had terrifying near-death experiences of their own or who had lost loved ones, the road to personal renewal was longer and more difficult. Even today, some Camille survivors struggle with intrusive distressing memories, recurring dreams of the storm, feelings of detachment, an inability to feel close to people, or an exaggerated startle response. Some are plagued by obsessive watchfulness, an overwhelming need for predictability (a dislike for surprises of any kind), or a desperate need for control.

There are some experiences in life we never forget, and memories of traumatic events like Camille can lay dormant in the subconscious, to be triggered without warning by certain smells or sounds. One survivor, on hearing the sound of a tornado on TV many years later, found himself mentally transported right back into the middle of the hurricane. When he became aware of his surroundings again, he was shaking and sweating. Memories of Camille are understandably triggered by heavy rains or lightning, but they have also been activated by more innocuous stimuli such as the sound of a helicopter, the sight of a refrigerated truck (in Nelson County, bodies were kept in such a truck pending identification), or, as with Luke Petrovich, the smell of sewerage.

Virtually every American was affected emotionally by the news of the 9/11 terrorist attacks, the 2004 tsunami in the Indian Ocean, and Hurricane Katrina in 2005, yet some of those who lived through Camille found that these media reports reactivated their own personal horrors from the past. Residents of Nelson County were especially affected emotionally by the prolonged search for victims of more recent disasters. Camille-era public officials express a particular sensitivity to the problems faced by community leaders in these emergencies, empathizing deeply with the overwhelming challenges of restoring infrastructure and services.

Some of Camille's survivors, remarkably, have never talked about the deaths of their loved ones in the disaster, even with their own families. Psychotherapists refer to such attempts to evade painful feelings as *emotional avoidance*. Others engage in *behavioral avoidance*, in which individuals shun situations that might stir up troubling memories. Luke Petrovich, for example, could not bring himself to see the movie *The Perfect Storm* because he'd heard there was a scene in which the water slowly rose over the heads of men in one boat until they drowned.

Sometimes the symptoms of mental distress are more severe. A study of the death rates in 377 U.S. counties in the 1980s found an appalling 31 percent increase in suicides in the three years following a major hurricane. A newspaper account of a Mississippi man's suicide several years after Camille suggested that he might legitimately be considered a casualty of the hurricane, given that he had been severely distressed since losing his family in that storm. (In the days immediately following Katrina, there were several reports of suicides in New Orleans, including two police officers.) Natives have described old friends as "never being right again" after Camille, and some, in moments of candor, even describe themselves in similar terms. But seeking therapy wasn't what ruggedly self-sufficient people did back in 1969. And the few exceptions weren't about to admit it to their friends, let alone encourage them to do the same.

Five years after Camille, the Disaster Relief Act of 1974 began to support certain mental health services in the wake of catastrophes. This federal legislation, however, did not immediately result in crisis intervention for disaster victims or rescue workers. It wasn't until the early 1990s that the Red Cross contracted with professional organizations such as the National Association of Social Workers and the American Psychological Association to provide such services after hurricanes and other disasters.

It became clear in the decades after Camille that family and social support networks can indeed mitigate the severe psychological impact of catastrophes. Camille survivors who talked about their experiences with family, friends, therapists, or even journalists years later, said it had helped. They did not become obsessed with it, but talking through the issues of trauma and loss seemed to have deprived the event of its power to overwhelm them, or to rear its ugly head in ways that undermined their ability to function in later life.

Talk therapy, however, presents special challenges when it comes to children. Youngsters cannot always *say* how they feel; instead, they are more likely to act out their feelings through anger, poor academic performance, or social withdrawal. Younger children in particular may mistakenly assume that the pandemonium around them is their fault. Yet parents, especially if they are under stress themselves, may wrongly assume that their children are resilient enough to quickly overcome upheavals in their young lives. The sooner children are returned to a normal routine after a disaster—attending school again, for instance—the quicker their recovery.

At the other end of the age spectrum, and despite their increased vulnerability to physical injury, elderly people are less likely to exhibit long-term psychological difficulties after a disaster. In Pass Christian, for example, Edith DeVries was well into her seventies and was recovering from a broken hip when Camille's storm surge swept her out of her collapsing house. She avoided drowning by hanging onto a plank, her head barely above the water, as the flood carried her well over a mile and deposited her in a grove of trees. In the wee hours of the morning, the receding water left her on the muddy ground, almost naked and too weak to even crawl. She lay there all day Monday, through Monday night, and most of the day Tuesday until National Guardsmen heard her feeble voice late that afternoon and took her to a hospital. Mrs. DeVries recovered, and her friends found her quite chipper, having apparently suffered little psychological distress as a result of her Camille experience. By the time a person has lived into their seventies and eighties, they have typically weathered considerable adversity and loss. They are simply more adept at taking things in stride.

Camille came and Camille went. In nature's grand scheme of things, the storm of the century was but a brief blip on the face of the globe. The planet still turns, rivers flow, tides rise and ebb. Gulls scavenge on the Mississippi beaches, alligators thrive in the Plaquemines bayous; even the scars on the Virginia mountains are gradually healing. Camille had few long-term effects on the geographical features of the stricken regions, nor did the storm affect the survival of any species. The disaster did, however, have another kind of legacy: It exposed the hollow illusion of individual and community self-sufficiency in the face of nature run amuck.

On the heels of Camille, federal flood insurance, which had been recently authorized by Congress but never resulted in the issuance of a single policy, quickly became an aggressive nationwide priority. Federal emergency response regulations were rewritten to eliminate bureaucratic biases against rural regions, and SBA loans and grants were made available even to victims who did not own businesses. Funding for the NWS was increased, permitting more rural field stations to collect and report data around the clock. These and related changes came in response to the demands of leaders in the victimized communities—the same locals who had until recently voiced stalwart opposition to anything that smacked of "big government."

Congress also began to realize that, in a country the size of the United States, disasters should be *expected* to occur fairly often. It was a virtual certainty that each year numerous communities would need federal assistance to restore damaged infrastructure: road and bridge repairs, the rechanneling of streams, removal of debris, and the like. After Camille, nobody could disagree about the government's responsibility to be continuously prepared for such tasks. In 1979, President Jimmy Carter created FEMA, not to manage disasters at the local level, but rather to coordinate the increased number of federal programs related to emergency response and disaster recovery. In 1996, President Bill Clinton elevated the agency to cabinet status in an effort to further improve its flexibility and power to respond in a timely manner. FEMA's responsibilities also grew to include not only reactive responses but also proactive disaster education for local response teams and the general public.

But bureaucratic memory can be short-lived, and in the wake of the World Trade Center attacks of 2001, FEMA was demoted from an independent agency to just another of the approximately twenty-three agencies under the newly created Department of Homeland Security. The first significant test of this new arrangement came with Hurricane Katrina's landfall in August of 2005.

As far as tropical meteorologists were concerned, the new technologies of satellite telemetry and computer storm-surge simulation had proven their meddle during Camille. With the question of feasibility resolved, tremendous strides were made in these and other analytical tools in ensuing years. Social scientists, meanwhile, found more federal and state funding available for studies of the human dimension of natural disasters. Their research findings gradually filtered into emergency management practices, including mental health interventions. Ultimately, all but those victims most determined to go it alone became more receptive to seeking mental health treatment after a disaster.

The new Saffir-Simpson Scale fulfilled its promise of facilitating local evacuation decisions. And it also did something more: as Her-

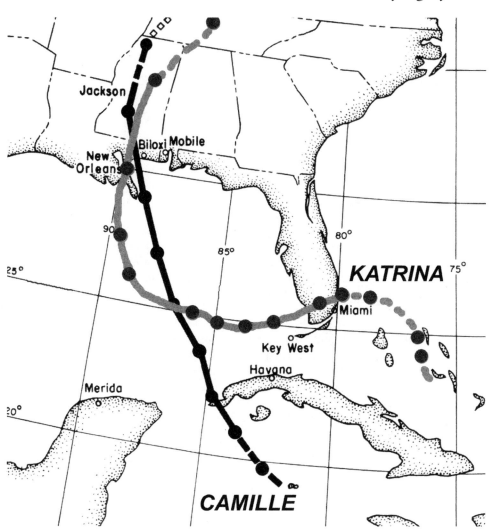

Comparative stormtracks of Camille, 1969, and Katrina, 2005.

bert Saffir had hoped, it raised the awareness of local building authorities in almost every hurricane-prone region of the country. Within a
few years after Camille, many local coastal governments stiffened
their building codes; some that had previously been lax even began to
enforce those codes. Today, the availability of federal flood insurance
is contingent on the local enactment of appropriate construction and

flood elevation standards—another encroachment of Washington into local affairs that very few people ever find reason to complain about.

Since Camille—and largely because of the hard lessons of that catastrophe—victims in stricken communities have come to expect no less than direct federal involvement in disaster response and recovery. Not that there is, or ever will be, a catastrophe without chaos; chaos is inherent in the nature of such events. But if there is ever a time when the fiercest of government bashers will welcome federal officials into their community, it will probably be on the heels of a natural disaster.

Just as this book was about to go to press, a Category 4 hurricane named Katrina roared in from the Gulf, and—in a replay of the Camille disaster—laid waste to lower Plaquemines Parish, parts of St. Bernard Parish, and the Mississippi Gulf Coast. As Katrina's storm clouds drifted off and the sun broke through, the city of New Orleans once again seemed to have been spared. That extraordinarily good luck, however, was short-lived; a few hours later, seawalls along several canals failed, the waters of Lake Pontchartrain poured through the breaches, and some 80 percent of the Crescent City was inundated.

Emergency responses to Katrina were uncoordinated, slow, and—at least in the early days—woefully inadequate. Politicians argued about whether there had been one disaster or two, as if that mattered. And before the last survivors were even evacuated, a flurry of finger-pointing had begun. But one question was conspicuously absent from the rhetoric: What have we learned from history?

Although it will be a year or more before the full costs of the recent catastrophe are tallied, it is already clear that Katrina was the most expensive disaster in American history, and by a considerable margin. And while it is too early to assign blame for the failures of foresight that aggravated the Katrina disaster, it's not too soon for planners and ordinary citizens to revisit Camille's earlier lessons. Mother Nature is under no obligation to wait another thirty-six years before sending along another event of Camille's violence. She could easily do so at any time.

Epilogue

A hundred miles north of Camille's landfall, ten-year-old evacuee Richard Rose awoke to a brilliant Monday morning, the air sweet with Orleans jasmine. As he dashed outside to meet the neighborhood kids, and his eighteen-year-old brother Don headed to Gulfport to check on their dad, the rest of the family stewed indoors in a brew of mounting worry. Late that day, a call came in on a police line. Richard will never forget his sister's piercing shriek. "Daddy's dead!"

The days that followed merged into a blur. The funeral home, the closed casket, Mother mistakenly spraying her hair with underarm deodorant, the mess at their Gulfport home, the chaos at Dad's business, their temporary stay with relatives who had a generator. Although nobody explained much to Richard about what had happened, eventually someone did admit to him that his dog Mickey had drowned in the garage.

Even as Richard grew older, the Rose family didn't discuss the tragedy. It was years later that he learned that his dad's body had been found naked in a chimney. Don had identified him by a ring and a scar from a bullet wound sustained in World War II.

In 1989, at age twenty-nine, Richard was elected to the Gulfport City Council—the youngest person ever to serve on that body. In 1997, toward the end of his fourth term, he undertook a reevaluation of his life. He enrolled in graduate school, earned a master's degree in public administration, and changed careers. For the next three years he worked on a FEMA project while spearheading an effort to obtain funding for a memorial to the thirteen people from Gulfport, including his dad, who died during Camille.

In 2002, Richard was invited to speak at the dedication of another Camille memorial—this time in Biloxi. He told the audience,

> As a ten year old, I stood near my father's closed coffin at Lang
> Funeral Home in Gulfport. I asked myself a question whose
> answer I believed I knew, but could not confirm . . . Is there
> any reason anyone should die in a hurricane?

In July 2004, he accepted the city manager's position in D'Iberville, a town on the north shore of the Back Bay of Biloxi. His first priority after taking on those responsibilities was to develop a hurricane response plan.

Barely a year later, the wisdom of Richard's foresight was affirmed by Hurricane Katrina. Although portions of the emergency plan were still a work-in-progress, the evacuation went smoothly, city hall was boarded up and became functional again shortly after the storm, and utilities and city water were soon restored. Unfortunately, 60 percent of D'Iberville's homes were destroyed, and even Richard's own home, although located five miles upstream along the Biloxi River and perched on pilings fourteen feet high, was flooded to a depth of five feet. The part of the new emergency plan that turned out to be weakest was the expectation of prompt federal aid. A full three weeks after the disaster, the feds had still not provided any temporary housing assistance for the local victims.

Mary Ann Gerlach's copious interviews with news reporters catapulted her tragic tale of the hurricane party into an urban legend. She made news again in January 1981, when she was arrested for murder. Although there were no eyewitnesses and the evidence was purely circumstantial, the state contended that Mary Ann had pumped five shots from a .357 magnum pistol into her thirteenth ex-husband, Lawrence A. Keitzer.

Mary Ann steadfastly insisted that she remembered nothing at all about the shooting. Her legal defense was straightforward: Her ordeal in surviving the destruction of the Richelieu during Hurricane Camille had rendered her legally insane. For good measure, her attorney tacked on a second somewhat contradictory plea: self-defense. On January 28, 1982, following an eight-day trial, a Harrison County jury found Mary Ann Gerlach guilty of murder.

It wasn't that the jury was insensitive to the long-term effects of disaster-related trauma; after all, each of them had also been affected by Camille's horrors. To accept a defense of inner turmoil and confusion as a legitimate excuse for sociopathic behavior, however, the jury would have likewise been excusing Camille's countless other walking wounded who still inhabited the Gulf Coast from adhering to standards of responsible behavior. How could any member of the jury do such a thing without abandoning all confidence in the continued recovery of their society? In the final analysis, they couldn't.

Mary Ann was sentenced to life imprisonment and placed in the custody of the Mississippi Department of Corrections. Her attorneys filed a motion for a new trial, partly on the basis of jury tampering, but it was denied. Then, from prison, Mary Ann wrote a flurry of letters to the Mississippi Supreme Court. The justices agreed to hear her appeal.

In the final ruling, which came down on March 20, 1986, the high court noted that Mary Ann had been "a regular victim of wife-beating." It went on to acknowledge a laundry list of erratic behaviors. The justices nevertheless denied her appeal, stating, "Just because you are crazy does not mean you are legally insane." Mary Ann was to serve out her sentence.

A sentence of life imprisonment, however, sometimes turns out to be less than that, and such was the case with Mary Ann. After an initial troublesome period of adjustment to incarceration, she settled into a level of resignation to her fate, became a regular churchgoer, took college courses and earned a liberal arts degree, and was characterized by the warden as a model prisoner. She married twice again while in prison, bringing her total to fifteen marriages by the time she was paroled in 1992.

Ben Duckworth, concerned that the parents of Rick and Luane Keller would think that their children had been reveling when Luane died, made every attempt possible to set the record straight about the infamous "hurricane party." Alas, although he was promised that his own version of those events would go out over the news wire services, Ben's rectifications actually got no farther than a local paper in Memphis. As for Rick Keller, he refused interviews about his experience in Camille and the loss of his beautiful young bride.

Frequent flashbacks and nightmares have haunted Ben. He made a trip back to the site of the Richelieu a year after Camille, but it didn't help. Ten years later, still having flashbacks and nightmares, he returned a second time. As he was about to hop across a gap in the concrete that had once been a stairwell, he was startled by a cotton-mouth moccasin basking in the sun. The snake and Ben stared at each other, and after a minute the reptile slithered away. After that the flashbacks ended. The nightmares, however, recurred from time to time.

Ben's last trip to the coast came at the invitation of Julia Guice, the former civil defense coordinator, on the occasion of one of the Camille anniversaries. Although Ben accepted reluctantly, after the ceremony he found himself compelled to visit the ruins of the Richelieu once again. This time, he had trouble finding the place, and only after hunting awhile did he realize that the site had become a shopping center. The "sad old hole" that had been the swimming pool was gone, and the last physical remnants of the defining event in his life had finally been erased. Within a week or so, for the first time in almost twenty years, Ben stopped having his nightmares.

The Mississippi coast took an economic nosedive after Camille. Hundreds of empty lots remained vacant and overgrown for as long as two or three decades, their sidewalks eerily leading nowhere. Although new, more stringent, building codes were quickly enacted, the goal of reducing the area's vulnerability to future hurricanes was undermined by the dire economic realities. Most residents couldn't even afford to rebuild to the former lax structural standards, let alone to an improved standard. One after another noncompliant permit application was approved as a special exception, and soon the improved building code became a paper tiger. The Gulf Coast began to look shabbier and shabbier. The completion of Interstate 10, rather than helping the coastal economy, served to route more traffic and commerce *away* from the smitten coast.

Then in 1990, the Mississippi State Legislature took the desperate action of legalizing gambling in the coastal counties. Incredibly, the new statutes specified that all gambling had to be done on the water—the very water that just two decades earlier had experienced the highest storm surge on record in the entire nation. One after another floating casino was built along the Gulf Coast, each outdoing the previous one in size and opulence. A spur was constructed linking

Interstate 10 with Biloxi Beach, and recreational gamblers poured in from across the southeast. When a hurricane warning would close the casinos, the state and local governments lost nearly five hundred thousand in taxes per day, and fifteen thousand casino workers went without paychecks. Yet casino construction and appurtenant coastal development continued unabated.

By the time Katrina struck in 2005, the towns along this shore were merged into a near-continuous sprawl of homes, apartments, businesses, and tourist attractions. While Biloxi grew modestly, from forty-four thousand to fifty-one thousand, Gulfport became a city of seventy-one thousand—up from thirty thousand in 1969. Memories of Camille, however, still lingered. Historical markers along the service road identified the sites of vanished architectural treasures like the Dixie White House and Trinity Episcopal Church. And a for-sale sign posted on an interior lot included the claim, "No flooding in Camille."

Meanwhile, Mississippi began to come to grips with its sordid history of racial bigotry. In 1994, Byron De La Beckwith was finally convicted of the 1963 murder of Medgar Evers. In 2005, former Klansman Edgar Ray Killen was convicted of manslaughter in the 1964 deaths of three young civil rights workers. This latter call for justice, albeit long delayed, resulted not only from the efforts of local grass root activists but also through the unflagging involvement of former Mississippi governor William F. Winter, now in his eighties, whose racial views stood in stark contrast to those of earlier Mississippi governors Ross Barnett and John Bell Williams. Even after Killen's conviction, however, Winter reminded audiences to not be too smug that all racial issues have now been resolved:

> I must tell you that the problem of race, despite all the progress that we have made, remains the thorniest, trickiest and most difficult barrier we confront to achieve a truly successful and united region. Most white folks think that we have come a lot further in race relations than most black people do. . . . It's a matter of developing a sense of trust. That is admittedly harder for blacks to do than for whites. For black people have more to forgive, even if they cannot and probably should not forget. But there must come a time when we have to recognize that we

are all in this together—when we must move past the old divisions of race and understand our common interests and our common humanity.

In Boothville, Louisiana, the Jurjevich family lost two consecutive homes to hurricanes in a span of just four years. Their insurance company paid them ten cents on the dollar in the Betsy disaster, and in Camille they had no homeowners insurance at all. But when HUD delivered a mobile home to his lot in late 1969, Leander Jurjevich had already built a high sturdy foundation for it. The possibility of moving out of Plaquemines Parish, or even out of Boothville, wasn't even considered.

To get the dirt for restoring the back levee after Camille, the Army Corps of Engineers scooped out a canal along the rear of the properties belonging to Jurjevich and his neighbors. As soon as the family was back on its feet financially, Leander built a stout private bridge across it. He wasn't about to be deprived of the grand view one gets from atop that levee.

A later addition was a big steel-sided barn. Inside, amid a collection of tools, his son Lea Jr. proudly showed off an airboat he skillfully built for himself. By all accounts, the vehicle was a Ferrari of the swamps. Next to that impressive piece of machinery was an even more astonishing sight. There sat the *Buffalo*, resting proud, spotless, and freshly painted from bow to stern. Through everything that had happened to them, the Jurjeviches held onto the old boat. It had now been in the family for almost ninety years, and it looked as river worthy as ever.

Then, in 2005, this region took a blow from Katrina, the levees failed once more, and as of this writing, it is still underwater.

It was sometime in 1972, with the recovery from Camille approaching completion, that Luke Petrovich noticed he was having a harder time getting out of bed in the morning. There was nothing wrong with him physically; he was simply mentally exhausted. Like many people who exercise leadership under such circumstances, he received no more than scant recognition for his efforts, neither in having saved the lives of thousands nor in having worked so tirelessly to bring back some semblance of normality to the birdsfoot delta. Luke realized it was time for him to have a personal life. He courted a woman twenty years his junior, married her in 1977, and at the age of fifty, started a family.

Meanwhile, Chalin Perez, president of the parish commission council, took to following the tradition of his late father, "the Judge," in viewing the parish as his personal fiefdom. He siphoned off parish mineral royalties for his family, diverted parish workers to his personal projects, and used the parish-owned helicopter for personal flights, including one in which the only passenger was his cocker spaniel. Then, as if there weren't enough public funds to satisfy the entire family's greed, Chalin Perez and his brother, Leander Jr., began bickering over who should get which slice of the parish pie.

By 1979, Luke had had enough. One day something triggered an image buried deep in the recesses of his mind a decade earlier—the sight of the sun rising behind the wreckage of the church steeple on the levee. Maybe it was a whiff of sewerage that brought back memories of that traumatic experience, or maybe it was something else. Whatever it was, Luke decided to take on Chalin and Lea Perez. The result was a long, bitter, and expensive legal war that eventually led to death threats against Luke's family. Luke, previously so stalwart against federal meddling, brought in the FBI and even grew to like those fellows.

One unanticipated consequence of Luke's battle against the Perez machine was that a cadre of blacks fought back for the first time. Calling themselves the Fishermans Association and Concerned Citizens of Plaquemines Parish, they protested the fact that the parish refused to apply for federal antipoverty funds such as Medicaid. They complained that Chalin Perez had used parish money to build a country club with a golf course and swimming pool that blacks weren't permitted to enter. And they got nationwide publicity for Chalin's claim that the parish, despite its hefty mineral rights, couldn't afford the three hundred thousand dollars required to bring public water lines to the blacks in the village of Ironton.

Luke eventually prevailed in the political struggle. The irony seemed lost on him that he did so with the assistance of the FBI, local blacks, and the outside news media—all groups he had earlier shunned. The Perezes were forced to return millions of dollars to the parish coffers, and their political base disintegrated as Luke rode a tide of popularity. After a government reorganization, Luke Petro-vich became the first president to be elected parishwide, and he was reelected for a second term. He went on to practice law in New Orleans until three weeks before he died in 2004 at the age of seventy-four.

With most of Nelson County's creeks and streams out of their nat-ural beds, Cliff Wood figured that the highest priority was to recover the farmland so essential to many folks' livelihoods. Rather than allowing the Army Corps of Engineers to quickly rechannel the streams, however, several groups of environmentalists insisted that impact studies be conducted, which annoyed Cliff to no end. To trump its shaggy-haired opposition, the corps identified one wayward stream after another as a hazard to public safety and therefore as an emergency situation that couldn't wait for environmental studies.

In talking about that unsettling fall of 1969, Cliff still has a ten-dency to apologize for not having done everything perfectly. He remains introspective about how things were handled after Camille, and he continues to second-guess himself even today. Cliff and his wife, Louise—"Ease" as he calls her—continue to live at Arrowhead, their home overlooking the James River, as they have for over forty years.

Sheriff Bill Whitehead lost his bid for reelection in 1971, as did many other longtime officeholders in Nelson County that year. Neither was Governor Godwin reelected (although he did succeed four years later). To Whitehead, this was a bitter blow—a slap in the face by the county folk he had always cared about so deeply. Cliff Wood was one friend who offered some consolation. Even Winston Churchill, Cliff reminded the sheriff, wasn't reelected in 1945. And without *him*, the British would be speaking German today.

What does a booted sheriff do next? Whitehead had his farm, but he was far too gregarious to make that his whole life. In 1972, he went to work for the state's Office of Civil Defense. Arriving there three years after Camille, he was astounded that the organization was still focused on a possible nuclear attack by the Soviets and that "not a damn soul was set up to deal with natural disasters." For the next fifteen years, he worked feverishly to try to enlighten that organization.

In November 2002, Bill was seriously injured and his wife Catherine was killed in an automobile accident. Wheelchair-bound and looked in on by a steady stream of family and friends, Sheriff Bill Whitehead lived out his remaining years at his beloved Willowbrook. He died in 2005 at the age of seventy-nine.

A couple of days after the storm, the bodies of Warren Raines's father, his eighteen-year-old sister Johanna, and his younger brother Sandy were found within three hundred yards of each other. A week later, his mother's body was discovered in the wreckage of a barn two and a half miles downstream. His younger sister Ginger was recovered a week or so after that, some eighteen miles downriver from Massies Mill and just two miles shy of being swept into the James River, which would have probably carried her all the way into Chesapeake Bay.

Warren and Carl's grandparents were too elderly and in no condition to keep two teenage boys. Their father's employer, Jack Young, encouraged the brothers to enroll in boarding schools because that would solve the problem of keeping a roof over their heads and food on the table. Young told them that people around the county had donated money for their tuition, and that time was of the essence since classes were to start in a few weeks. The boys didn't like the idea, but they felt they had no choice. A few years later, they would discover that their tuition had actually been paid with proceeds from their father's life insurance.

Warren hastily chose a now-defunct military school in Staunton, Virginia, so he could be near Mr. Young, whom he regarded as a substitute father. Only when he arrived on campus did he learn that the school was intended for boys with disciplinary problems. Warren's life went from one nightmare to another, and in his words, "it was hell on earth." If a student misbehaved in any way, the teacher paraded him to the front of the class and had him bend over and grab his ankles. After getting five whacks on the behind, no boy could possibly return to his desk without tears in his eyes—a double humiliation. Sometimes the entire student body was punished for the misdeeds of one person; once, they were required to stand at attention outside in the cold rain from 9:30 p.m. until 2:30 a.m., until someone finally confessed. The school buildings, dating from before the Civil War, did not make for comfortable living. No emotional support was forthcoming from the staff, which was more accustomed to dealing with unruly youngsters than with a grieving adolescent. Warren had no one to turn to, no one to talk to, not even his dog, Bo. He was starved for affection.

While Warren spent that miserable year in military school, Carl Jr. lived with a family that had offered to take the boys—three weeks too late for Warren, whose hefty tuition had already been paid. At the end of that school year, Warren moved in sequentially with two different families, then he rented a room from a widower for a few months, and finally—with his brother Carl now in college—he returned to Massies Mill alone. The Raines home had stood vacant for two and a half years, the mud having been cleaned out and the windows replaced by the Mennonite Disaster Service. There was still furniture on the second floor, and as difficult as it was to be there alone, at least it was home.

The last time he'd lived there, it had been with six other family members. At first he had pleasant dreams of his family being alive, only to wake up and find himself confronted with the dark and silent reality. Then the nightmares resumed.

In the summer of 1973, he met a young woman named Sharon White at church services. She turned out to be a niece of Tommy Huffman, who had lost so many of his own kin in the flooding on Davis Creek. She was wise beyond her years in talking about tragedy, sorrow, and redemption. In April 1974, when Warren was nineteen, he and Sharon were married. They still are.

Warren has every letter anyone ever wrote to him while he endured his year of misery in Staunton. His grandmother wrote things like, "Son, we're gonna get you out of there. We're doing all we can to get you out." Warren doesn't read them any more; they upset him too much—yet he keeps them. And of course he'll never forget Camille. Today, if a helicopter flies overhead, for Warren it's like pushing the play button on a VCR; that sound brings it all back.

The evening after the disaster, Tommy and Adelaide Huffman prepared food from their thawing freezer on an outside fire pit and shared it with seventeen relatives and homeless neighbors. For the next several months, their basement was wall-to-wall with mattresses accommodating various family members. Yet the thought of asking the Red Cross for help in feeding everyone never occurred to the Huffmans; that would have been begging, and they were not a begging kind of people.

No sooner had one lane of Highway 29 been opened at Woods Mill when a Sears truck from Charlottesville rolled into the Huffmans' driveway. Its contents: a new gas range to replace the old electric one. Adelaide still relates how much that range helped, considering that it would be nearly a month before electricity was restored. They tried to find out who had sent it so they could pay him back, but to no avail. The benefactor, whoever it was, remains anonymous to this day.

It was growing apparent that anyone who hadn't been found already had not survived, and the only news the Huffmans could expect was that the body of one or another of their missing kin had been found somewhere. Brothers Tommy and Russell made daily trips to the county's makeshift morgue—a refrigerated tractor trailer set up next to the funeral home—to examine and identify the bodies that kept arriving. There were so many of the Huffmans missing that a phone company crew strung a dedicated line between the funeral home and the single surviving Huffman home. For more than four weeks, Tommy or Russell made that sad trip daily, usually returning to the family homestead with news that brought closure to the fate of yet another of their family members or neighbors. Six of the twenty-two of their kin who died in Camille, however, would never be found.

If it's possible for one death to be more tragic than another, that judgment might apply to Tommy's brother Lawrence. Only a few days before the disaster, his arm was severed in a workplace accident. Had his hospital treatment been extended just one more day, he would have missed the flood. As it turned out, he returned home on the day the rains came, and his amputation was the characteristic used to identify his body.

The only positive surprise during the whole traumatic affair came with a knock on the Huffman door on the evening of Day 3. There stood Tommy's brother James, whose home and entire family were gone. Unbeknownst to the rest of the surviving Huffmans, James had volunteered to work a night shift in the next county, and it had taken him three days to get back. He'd had to walk the last few miles from the road blockage at Woods Mill. Even today, when Tommy speaks of seeing his brother who had been given up for dead, his voice still cracks with emotion.

Tommy and Adelaide Huffman live in the same neat, modest brick home that, working weekends, they built with their own hands four decades ago. He still keeps cows on the mountainsides of Huffman's Hollow. Davis Creek is hidden from view in the woods below their yard, but from their side porch you can glimpse the church and the cemetery where so many of their loved ones were laid to rest—including their son who survived Camille but drowned years later at age thirty.

The Huffmans' roots run deep in these hills and hollows. They speak slowly and quietly, drawing from some invisible yet solid source of inner strength. They patiently answer questions, and expect nothing in return for their help—nothing except fairness in what is written about the night that changed their world.

Nelson County still has no industry to speak of; the small American Cyanamide plant has long since closed, and a soapstone company where Earl Hamner Sr. was once employed has only a small labor force. Most of the natives are involved in agriculture, mainly apple orchards or corn, or they commute to jobs outside of the county. A resort community has been developed on a fourteen-thousand-acre tract of mostly high ground, and it not only doubled the tax base almost overnight, but is now the county's largest employer. The population of this picturesque county in the Blue Ridge foothills has grown, but only slightly, from about 12,000 in 1969 to the present 14,500.

Today's visitor to Nelson County does not see much evidence of the tragedy that befell these rural folk that fateful night in 1969. People go about their lives. The Tye and Rockfish Rivers are behaving properly as scenic little streams. The land, like the people, has mostly recovered, the mountainsides having reclaimed their verdant cover. But in some places bare rock is still visible on the mountainsides. Almost hidden, scars remain.

The Saffir-Simpson Hurricane Potential Damage Scale

TROPICAL DISTURBANCE
No strong winds. Rotary circulation is apparent in satellite photographs, although it may be slight or absent at the surface. Isobars (lines of equal barometric pressure) are discontinuous. A common phenomenon in the tropics.

TROPICAL DEPRESSION
Sustained (one-minute) winds do not exceed 39 mph. Some rotary circulation extends down to the surface. At least one isobar forms a continuous closed loop.

TROPICAL STORM
Sustained wind speeds between 39 and 74 mph. Distinct rotary circulation at all altitudes. Closed isobars.

CATEGORY 1 HURRICANE (MINIMAL DAMAGE)
Wind speeds between 74 and 95 mph. Strong and very pronounced rotary circulation. Central barometric pressure typically 28.94 inches or higher. Possible storm surge of four to five feet. Minimal damage, mainly to trees, shrubbery, unanchored outbuildings, mobile homes, and poorly constructed signs. Possible coastal road flooding and minor damage to docks and piers. Possible interruption of utilities (electric, gas, water, and landline phone service).

CATEGORY 2 HURRICANE (MODERATE DAMAGE)
Wind speeds between 96 and 110 mph. Barometric pressure between 28.50 and 28.91 inches. Possible storm surge of six to eight feet. Some roofing, door, and window damage to buildings. Considerable damage to shrubbery and trees with some trees blown down; considerable damage to mobile homes, poorly constructed signs, and piers. Coastal and low-lying escape routes may flood two to four hours before arrival of the hurricane center. Small craft in unprotected waters break moorings. Likely loss of utilities.

CATEGORY 3 HURRICANE (EXTENSIVE DAMAGE)
Wind speeds between 111 and 130 mph. Barometric pressure between 27.91 and 28.47 inches. Possible storm surge of nine to twelve feet. Some structural dam-

age to residences and utility buildings, occasional curtain-wall failures. Some large trees uprooted, mobile homes destroyed. Low-lying escape routes may be cut off by rising water three to five hours before arrival of the center of the hurricane. Flooding near the coast destroys smaller structures, with larger structures damaged by battering from floating debris. Terrain continuously lower than five feet above mean sea level may be flooded inland eight miles or more. Significant damage to small boats and parked aircraft. Certain loss of utilities for an extended period over a widespread region. Some roads impassible for several days.

CATEGORY 4 HURRICANE (EXTREME DAMAGE)

Wind speeds between 131 and 155 mph. Barometric pressure between 27.17 and 27.88 inches. Possible storm surge of thirteen to eighteen feet. Extensive curtain-wall damage to buildings; some complete roof failures. Extensive damage to windows and doors. Low-lying escape routes are cut off by rising water three to five hours before arrival of the center of the hurricane. Major flood damage to structures near shore. Residential areas at an elevation of ten feet or less within six miles of the coast may be flooded and should be evacuated. Damage to low-lying roads and bridges. Numerous pleasure craft destroyed even in protected waters. Parked aircraft destroyed. Serious damage to utility infrastructures.

CATEGORY 5 HURRICANE (CATASTROPHIC DAMAGE)

Sustained winds in excess of 155 mph. Barometric pressure less than 27.17 inches. Possible storm surge exceeding eighteen feet. Complete roof failure on many residences and industrial buldings; numerous complete building failures with smaller unanchored structures blown down or away. Most buildings within a quarter mile of shore totally destroyed; extensive erosion damage to roads; docks and low-lying bridges swept away; large ships and barges grounded. Large segments of utility infrastructure totally destroyed.

Category 4 and 5 Hurricanes Making Landfall in the United States since 1900

Name	Date	Category at Landfall	Sustained Winds[a] (mph)	Landfall
Camille	Aug. 1969	5	172	MS, LA
unnamed	Sept. 1935	5	160	FL Keys
Andrew	Aug. 1992	5[b]	155	south FL, LA, MS
Katrina	Aug. 2005	4	145	south FL, LA, MS, AL
Charley	Aug. 2004	4	145	central FL
Carla	Sept. 1961	4	145	north TX
Donna	Sept. 1960	4	145	FL to New England
unnamed	Sept. 1928	4	145	central FL
unnamed	Sept. 1919	4	145	FL Keys, TX
unnamed	Sept. 1915	4	145	LA
unnamed	Sept. 1909	4	145	LA
unnamed	Sept. 1900	4	145	TX (Galveston)
Hugo	Sept. 1989	4	140	SC
Audrey	June 1957	4	135	west LA
Hazel	Oct. 1954	4	135	SC, NC
unnamed	Sept. 1947	4	135	FL, LA, MS
unnamed	Aug. 1932	4	135	TX
unnamed	Sept. 1926	4	135	FL, MS, AL
unnamed	Aug. 1915	4	135	TX (Galveston)

Source: National Oceanic and Atmospheric Administration.

[a]Wind speeds are approximate and as of landfall; some of these storms had higher wind speeds while offshore.

[b]Andrew was reclassified from Category 4 to Category 5 in 2002.

Notes on Sources

1. Grim News

The material on the Duckworth family is based on our personal interviews with Josephine and Ben, supplemented by a series of follow-up phone conversations and personal correspondences with Ben and his sister, Marian, and her husband, Bill. We drew additional details from photographs supplied by Ben and from our personal observations. The Budweiser beer trucks were shown in the video production "Camille: 30 Years After"; a similar instance in Plaquemines Parish was described by Buras (1995). The Walter Cronkite quotation is as related to us by Ben Duckworth; our several requests to CBS for a transcript to confirm the exact wording went unanswered.

Slightly different versions of Mary Ann Gerlach's story have found their way into print; the one we include here is based primarily on the transcript of Pyle's 1980 interview in the oral history collection at the University of Southern Mississippi Library in Hattiesburg. This version differs in some details from the earlier newspaper accounts. Several details, such as Fritz's inability to swim and the condition of his body, are based on the testimony of witnesses at Ms. Gerlach's January 1982 trial for the murder of her thirteenth ex-husband.

The material on the Raines family is based on interviews with Warren and Sharon Raines, on contemporary photographs, and on our personal observations in Nelson County. Background information on Nelson County can be found in Simpson and Simpson 1970 and Pollard 1997.

2. Of Love and Life

Pielke and Pielke (1997) discuss hurricane risks and settlement patterns. Information on the Cheniere Caminada and Galveston disasters, as well as the meteorological information on Hurricane Audrey and the crawfish account, is from National Oceanic and Atmospheric Administration (NOAA) records. Some related anecdotal information came from local residents, including Juanita Rougeau. Geographical descriptions are based on our firsthand observations, and some details are from Menard 1999. The Broussard family's story is based on our personal interview, supplemented by details in an unpublished manuscript written by their son Brent, who died in a plane crash.

A range of figures on deaths and damages from Hurricane Audrey has appeared in print; we quote the figures from those official sources that in our judgment are the most credible. Our figure of 556 fatalities includes the missing who were never found. Establishing any casualty figure, however, is fraught with epistemological difficulties, and the figure we give here differs from some of the other published tallies.

3. Bayou Country

Our geographical and cultural overview of Acadiana is based on a combination of personal observations and experiences, the writings of other outside visitors such as Tidwell (2003) and Jenkins (1995), and from the reflections of various natives. Much of the historical material is from Taylor 1984; Calhoun 2002; and Barry 1997. Criticisms of the "man over nature" philosophy as applied to the coasts are found in Pielke and Pielke 1997; Morgan 1971; Barry 1997; and a growing number of other studies.

Marc Levitan, director of the Hurricane Center at Louisiana State University, was particularly helpful in sharing his thoughts and experiences regarding issues of evacuation. Additional material relating to evacuation can be found in Baker 1991; Carter, Clark, and Leik 1979; Christenson and Ruch 1980; Drabeck and Stephenson 1971; Fischetti 2001; Perry 1979; Perry and Lindell 1991; and Weinstein 1989.

4. The Birdsfoot Peninsula

The descriptive material on Plaquemines Parish is based on our own observations and on our conversations with various natives. The 1927 dynamiting of the levee is described by Barry (1997). Conaway (1973) and Jeansonne (1977) also provided some of the background on the 1927 incident, on Luke Petrovich, and on Hurricane Betsy. Additional material is based on our personal interviews and follow-up phone conversations with Luke Petrovich and with Charmain, Lea Jr., and Leander Jurjevich.

We limited our treatment of Hurricane Betsy to what we felt necessary to provide background for the later chapters. Although Betsy reached Category 4 intensity out in the Gulf, her winds decreased to Category 3 status just as she struck land, and she may have even been down to Category 2 when she hit Venice and Buras. This mention is not intended to trivialize the human impact of this serious storm; in fact, flooding was severe. Our brief account of Betsy is based primarily on 1965 newspaper accounts in the New Orleans *Times-Picayune,* on tracking data and photographs from NOAA, and on Sullivan 1986.

Williams (1969) provides some of the general political background; the more specific material on Leander Perez is based on the biographies by Conaway (1973) and Jeansonne (1977). Supplemental detail came from Rather 1963.

We found a number of Plaquemines Parish natives, both black and white, reluctant to talk publicly about Perez; the comments they made to us "not for attribution," however, suggest that the shadow of Perez may continue to affect some of the politics of this region even today. As for the political environment when Hurricane Camille struck in 1969, there can be little doubt of its racially biased and oligarchic nature, and we vouch for the accuracy of the picture we have painted here.

5. Storm Warnings

The weather reports and hurricane advisories are quoted verbatim from NOAA archives. For readers who find themselves confused about the sometimes conflicting terminology of "Weather Bureau" and "Weather Service," we point out that the published sources were not particularly consistent in 1969. A year prior to Camille, the U.S. Weather Bureau was officially renamed the National Weather Service. Numerous documents, even many from government sources, nevertheless continued to use the term "Weather Bureau" for several years after the official name change.

Robert H. Simpson, who reviewed the early drafts of this material, kindly supplied many of the relevant facts in a series of communications and a phone interview. Some of the material on the Corpus Christi hurricane was also based on *Station Report of the Hurricane of 1919,* issued by Charles A. Heckathorn, meteorologist-in-charge of the Weather Bureau Office in Corpus Christi. The quotation by Norton is from Sheets and Williams 2001. The story of Edward R. Murrow is based on Anderson 1955. Other material came from Robert Simpson's own publications, including Simpson 1971a.

Including Dr. Simpson's personal thoughts and activities during and after Camille presented a challenge to us as writers, insofar as he was steadfastly unconvinced that anyone could possibly be interested in Simpson as a person rather than Simpson as a scientist. In talking with him by phone, however, and in exchanging e-mails with him (in the large type he requested), it became apparent to us that here was a thinker who had a deeply ingrained sense of his social responsibilities and that to ignore Simpson *the man* in this book would be a great disservice to the story of Camille. At the age of ninety, Dr. Simpson reviewed, corrected, and approved most of the material we include about him in this and later chapters. We assure the reader that we have not engaged in any wild speculations about Simpson's thoughts, actions, motivations, or retrospections. It was Simpson who mentioned the importance of Chester Jelesnianski's seminal work on predicting storm surges, whose technical aspects are documented in a variety of NOAA reports under Jelesnianski's authorship between the late 1960s and the early 1990s.

6. On the Coast

Descriptions of Luke Petrovich's activities on August 16–17 are based on our personal interviews and are generally consistent with the U.S. Army Corps of Engineers reports. The materials on the history and culture of the Mississippi Gulf Coast and of racial tensions in the region are from Mason, with Smith, 2000; Applebome 1996; Horton and Horton 2001; La Violette 2001; Rubin 2002; and Scharff 1999, supplemented by our conversations with white and black residents of the region. The racial climate in Mississippi is also examined by Moody (1968) and Borstelmann (2001). Information on the Gulfside Assembly is based on a personal visit and an interview with the director, Ms. Marian Martin.

Most of the statistics are from the Bureau of the Census, U.S. Department of Commerce. The material on Wade and Julia Guice is based on transcripts of their 1980 interviews in the oral history collection at the University of Southern Mississippi. The general substance of the pre-landfall executive committee meeting was reported in the Biloxi *Daily Herald* on Sunday, August 17, 1969; we fleshed out this description based on various snippets in the other aforementioned sources. Greg Durrschmidt's account is based on DeAngelis 1969, supplemented by our telephone interview with Durrschmidt. General background material is drawn from photographs in the archives at the Biloxi Public Library and our direct observations of the region. The historical pattern of regional hurricanes as well as the tracking data on Camille are from NOAA archives.

7. Exodus

The material on Ben Duckworth and Richard Rose is from our personal interviews. The media issues are examined by Burkhart (1991). The series of hurricane advisories and warnings is from NOAA. Evacuation statistics and descriptions of traffic snarls are from the U.S. Army Corps of Engineers reports, as are the Corps's own preparations.

Robert Clark's quote is from Sheets and Williams (2001, 152–53). The scientific principles are documented in a variety of college-level textbooks, including those written by Zebrowski.

The description of the "Trappers War" is based on Jeansonne 1977. The general description of Delacroix Island is based on our firsthand observations, while the effects of Hurricane Betsy on that village are from our conversations with local shrimpers. Evacuation behaviors are consolidated from Wilkinson and Ross 1970; Perry 1979; and Baker 1991. Other hurricane preparations, including those in Mobile and Pensacola, are as reported in the New Orleans *Times-Picayune* in the days immediately preceding and following the disaster.

Lucretia and Sally's horror story was published in a New Orleans newspaper in 1969; additional details are from personal correspondence and a subsequent conversation with Lucretia. The women's names have been changed here to protect their privacy.

The story of the *Rum Runner* is from our interview with Ronald Durr; an article in the *Times Picayune* on August 20, 1969; and Shrake 1970.

8. Troubled Waters

The background material on river pilots is based on a series of feature articles in the New Orleans *Times Picayune*, November 4–7, 2001; an article by John Maginnis in the Ruston *Morning Paper*, June 4, 2004; our own visit to Pilot Town, and our phone interview with Paul Vogt Sr. Additional information was obtained through written correspondence with Jeff Bowdoin, Coast Guard historian, and phone conversations with officials at the Louisiana Department of Transportation.

The quotation on marine losses is from USACE 1970b, 44. Other information on the storm is from the same source. Material on the *Buffalo* and the family's hurricane experiences is based on interviews and follow-up phone conversations with Leander, Charmain, and Lea Jr. Jurjevich, who were kind enough to show us the old boat. The accounts relating to Garden Island, the weather observatory, and the wind measurements reported by the *Cristobal* are from the Army Corps of Engineers reports. The experiences of the four men in the sewerage facility are from a personal interview and several follow-up phone conversations with Luke Petrovich. The continuation of Ben Duckworth's story is based on personal interviews.

9. Angry Seas

The continuation of the *Rum Runner* episode is based on a personal interview with Ronald Durr. Most of the other material in the chapter, including the ordeal of Fathers Murphy and Cavanaugh, is pieced together from articles appearing in newspapers, including the New Orleans *Times Picayune* and the Biloxi *Daily Herald* in the week following the disaster. The Williams family story is from Paul Williams's oral history at the University of Southern Mississippi. The meteorological and hydrological information is drawn from the post-disaster reports of the Army Corps of Engineers and from the other sources cited within the text. The continuation of Ben's story is based on our interviews with him and members of his family.

10. Dawn

The narratives in this chapter are based on our interviews with the individuals cited, on our personal observations, and on Mason, with Smith, 2000. Snippets are drawn from local newspaper accounts and an interview with Bobby and Missy Goff. Readers interested in racial behavioral differences after disasters may refer to Christenson and Ruch 1980; Peacock, Morrow, and Gladwin 1997; and Wilkinson and Ross 1970. Leander Perez's involvement in the battle of Biloxi Beach is discussed in Jeansonne 1977; the quote is from p. 316. Earlier, on p. 169, Jeansonne discusses Perez's influence on the thinking of then congressman and future governor John Bell Williams.

11. Rubble

Jackson M. Balch's account, including the direct quotes, are from his testimony at the U.S. Senate hearings (United States Senate 1970a, 379–400). Mississippi newspapers described the destruction along the coast, as did the USACE Mobile District report. The effects on shipping and the oil industry are from Rohlfs 1969 and from USACE 1970b. Although the U.S. Coast Guard does not maintain comprehensive records of shipping accidents unless the Coast Guard was directly involved, their historian, Jeff Bowdoin, was helpful in supplying several general documents. The Louisiana Department of Transportation advised us that they do not maintain records on shipping accidents in the river (although by law that department is apparently responsible for doing so). A representative of the Branch Pilots' Association also advised us that they do not keep such records. The U.S. Mineral Management Service did not respond to our requests for information about losses of offshore drilling rigs and platforms during Camille; that agency has, however, published reports relating to more recent disasters.

Anecdotal information about looting came from our conversations with Camille survivors. Our conclusions about looting are consistent with Drabeck and Stephenson 1971; Quarantelli 1994; and our discussions with John Pine, director of the Disaster Science Management concentration at Louisiana State University.

12. Deluge

Descriptions of the storm are based on NOAA archives; the photographic and oral history archives in the Nelson County Library; and personal interviews with Warren and Sharon Raines, Tommy and Adelaide Huffman, Sheriff Bill Whitehead, Cliff Wood, and other Nelson County natives. The quiet zone information came from the U.S. Senate hearings and from Bill Whitehead. Some of the descriptions are based on our personal observations as we drove and tramped through the affected areas. Pollard 1997 and Simpson and Simpson 1970 were helpful to us in filling in some of the details. Hydrological data is from the Army Corps of Engineers Norfolk District report.

13. A County Divided

This chapter draws from the same sources as chapter 12. The multiple accounts correlate remarkably well considering the circumstances, although occasionally their time scales do disagree slightly. On the few occasions where we encountered unresolvable discrepancies between the human stories, we omitted that material.

14. Reconnecting

Our observations about sociological effects are generally consistent with those of Charles Fritz's early work (1961) as well as other researchers, including Kreps (1989), Perry (1982), and Steinberg (2000). The remainder of this chapter is based on our personal interviews with survivors.

15. Outsiders

The problem of coordinating relief organizations is a thread running throughout the Senate hearing transcripts (United States Senate 1970a, 1970b, 1970c). The problems with ants and lost pets were described by numerous survivors and emergency workers and are mentioned in the USACE reports (1970a, 1970b, 1970c) and in articles appearing in the *Times-Picayune* in the days following the disaster. Thomas Regan's remark is from the Senate hearing transcript (United States Senate 1970c, 1219). Mason's quote about Red Cross discrimination is from his book (with Smith, 2000, 176). Discrimination in the postdisaster evacuation is described in Mason, with Smith, 2000; in the Senate hearing transcript (United States Senate 1970b); and was confirmed in several of our personal interviews. Robertson's testimony about his investigation is in United States Senate 1970c; the complete report is in the archives of the library of Washington and Lee University.

The description of the creation of the Saffir-Simpson Scale is based on our interview with Robert Simpson and on NOAA publications. The material on Herbert Saffir is from standard biographical sources, a conversation with Marc Levitan, one of his former proteges, and an interview by Ann Carter that appeared in the Miami *Sun-Sentinel* on June 24, 2001.

16. A Knotty Legacy

The information on "Faith, Hope, and Charity" is from the Biloxi *Sun Herald*, August 18, 2000. The discussion of the eight unclaimed bodies in Nelson County is based on an article by Meg Hibbert in the *Nelson County Times* on August 17, 1989, on the twentieth anniversary of Camille.

Differences in how sociologists and mental health professionals view disasters are based on Edwards 1998 and on our own observations. Further information on stress reactions can be found in American Psychiatric Association 1994. Information on suicide rates after natural disasters came from Krug et al. 1998. For further information on mental health services after disasters, refer to Lystad 1988. Children in disasters are discussed by Zubenko and Capozzoli (2002) and La Greca et al. (2002). The story of Edith DeVries came from her story as told to W. G. Fennell, in mid-September 1969, and as described in Olin Clark's oral history taken by Ms. Charlotte Capers in the University of Southern Mississippi recorded on August 28, 1969.

Social scientists recognize that poverty is a major variable in postdisaster recovery. For information on how socioeconomic status and other variables mitigate or aggravate recovery, see Bolin 1982; and La Greca et al. 2002. Perry and Mushkatel 1986 and Perry and Lindell 1991 examine issues of minorities in disasters. Are Holen's (1990) study of disaster survivors is recommended for detailed information on long-term psychological and physical consequences of trauma. The Red Cross acknowledges the need for psychotherapy for delayed-onset stress reactions after disasters and, according to a *New York Times* article on August 21, 2002, agreed to pay for such extended treatment for survivors after 9/11.

Bibliography

Libraries Visited
Biloxi Public Library, Biloxi, Mississippi
Lincoln Parish Library, Ruston, Louisiana
Louisiana State University, Baton Rouge, Louisiana
Louisiana Tech Library, Ruston, Louisiana
McCain Library, University of Southern Mississippi, Hattiesburg, Mississippi
Nelson County Library, Lovingston, Virginia
Plaquemines Parish Library, Buras, Louisiana
Southern University Library, Baton Rouge, Louisiana
Washington and Lee University Library, Lexington, Virginia

Published Sources
American Psychiatric Association. 1994. *Diagnostic and Statistical Manual of Mental Disorders.* 4th ed. Washington, DC.
Anderson, Col. William C. 1955. "Edward R. Murrow and the Eye of the Hurricane." *Retired Officer Magazine,* April, 55ff.
Applebome, Peter. 1996. *Dixie Rising: How the South Is Shaping American Values, Politics, and Culture.* New York: Harcourt Brace.
Aptekar, Lewis. N.d. "The Psychosocial Process of Adjusting to Natural Disasters." Institute of Behavioral Science, Natural Hazards Research and Applications Information Center, Working Paper #70. Boulder: University of Colorado.
Baker, E. J. 1991. "Hurricane Evacuation Behaviour." *International Journal of Mass Emergencies and Disasters* 9 (2): 287–310.
Balk, David. 1996. "Attachment and the Reactions of Bereaved College Students: A Longitudinal Study." In *Continuing Bonds: New Understandings of Grief,* ed. Dennis Klass, Phyllis Silverman, and Steven Nickman. Philadelphia: Taylor and Francis.
Barnes, Captain Waldemar F. 1969. "History of the 53rd Weather Reconnaissance Squadron at Ramey Air Force Base, Puerto Rico, 1 July 1969 to 30 September 1969." 9th Weather Reconnaissance Wing, Air Weather Service (MAC), United States Air Force. Typewritten document.
Barry, John M. 1997. *Rising Tide.* New York: Touchstone.
Bolin, Robert C. 1982. *Long-Term Family Recovery from Disaster.* Institute of

Behavioral Science, Program on Environment and Behavior, Monograph #36. Boulder: University of Colorado.

———. 1989. "Natural Disasters." In *Psychosocial Aspects of Disaster,* ed. R. Gist and B. Lubin, 3–28. New York: Wiley.

Bolin, Robert, and Paticia Bolton. 1986. *Race, Religion, and Ethnicity in Disaster Recovery.* Institute of Behavioral Science, Program on Environment and Behavior, Monograph #42. Boulder: University of Colorado.

Borstelmann, Thomas. 2001. *The Cold War and the Color Line.* Cambridge, MA: Harvard University Press.

Bourne, Joel K., Jr. 2004. "Gone With the Water." *National Geographic* 206 (4): 88–105.

Breaux, Raymond R. 1993. "Gulfside: Seventy Years of Service." *New World Outlook,* March–April, 16–19.

Broder, David S. 2005. "Mississippi Healing." *Washington Post,* January 16, B7.

Buchanan, William, et al. 1970. *The Hundred Year Flood: Reactions to Hurricane Camille in Nelson, Amherst, and Rockbridge Counties, Virginia.* Lexington, VA: Washington and Lee University.

Buras, Janice P. 1995. *Betsy and Camille: Sisters of Destruction.* Belle Chase, LA: Down the Road Publishing.

———. 1996. *Way Down Yonder in Plaquemines.* Gretna, LA: Pelican Publishing.

Burkhart, Ford N. 1991. *Media, Emergency Warnings, and Citizen Response.* Boulder: Westview Press.

Calhoun, Milburn, ed. 2002. *Louisiana Almanac.* Gretna, LA: Pelican Publishing.

Carter, T., J. Clark, and R. Leik. 1979. *Organizational and Household Response to Hurricane Warnings in the Local Community.* NHWS Report Series, Department of Sociology. Minneapolis: University of Minnesota.

Christenson, L., and C. Ruch. 1980. "The Effect of Social Influence on Response to Hurricane Warnings." *Disasters* 4:205–10.

Coast Guard News. 1969. News Release. New Orleans: Public Information Office, Customhouse. August 27.

Conaway, James. 1973. *Judge: The Life and Times of Leander Perez.* New York: Knopf.

Conway, Eric D., and Maryland Space Grant Consortium. 1997. *An Introduction to Satellite Image Interpretation.* Baltimore: Johns Hopkins University Press.

Darce, Keith, and Jeffrey Meitrodt. 2001. "Masters of the River." *New Orleans Times-Picayune,* November 4.

Davies, Pete. 2000. *Inside the Hurricane: Face to Face with Nature's Deadliest Storms.* New York: Henry Holt.

DeAngelis, R. M. 1969. "Enter Camille." *Weatherwise* 22 (5): 173–79.

Drabeck, T. W., and J. S. Stephenson. 1971. "When Disaster Strikes." *Journal of Applied Social Psychology* 1 (2): 187–203.

Edwards, Margie L. 1998. "An Interdisciplinary Perspective on Disasters and Stress: The Promise of an Ecological Framework." *Sociological Forum* 13 (1): 115–32.

Ehrbright, Nan Patton. 1989. "Woman Says Hurricane Led to 1981 Murder of Her Husband." *Biloxi Sun Herald*, August 13, 20–22.

Enarson, Elaine, and Betty Hearn Morrow, eds. 1998. *The Gendered Terrain of Disaster: Through Women's Eyes.* Westport, CT: Praeger.

ESSA (Environmental Science Services Administration). 1969a. *Hurricane Camille—A Report to the Administrator.* Washington, DC: U.S. Department of Commerce.

———. 1969b. *The Virginia Floods, August 19–22, 1969: A Report to the Administrator.* Washington, DC: U.S. Department of Commerce.

Everly, G. S., Jr. 2003. "The Evolving Nature of Disaster Mental Health Services." *International Journal of Emergency Mental Health* 5 (3): 113–19.

Fischetti, Mark. 2001. "Drowning New Orleans." *Scientific American,* October, 77–85.

Fritz, Charles E. 1961. "Disaster." In *Social Problems,* ed. Robert K. Merton and Robert Nisbet, 651–94. New York: Harcourt Brace & World.

———. 1996. "Disasters and Mental Health: Therapeutic Principles Drawn from Disaster Studies." *DRC Historical and Comparative Disaster Series* #10. Newark: Disaster Research Center, University of Delaware.

Gillette, Becky. 2002. "Big Gamble for the Coast Leads to Cinderella-Like Success." *Mississippi Business Journal,* August 19–25, 34.

Gleick, James. 1987. *Chaos: Making a New Science.* New York: Viking.

Godschalk, D. R., D. J. Brower, and T. Beatley. 1989. "Mitigation after Camille, Frederic, and Alice." In *Catastrophic Coastal Storms.* Durham, NC: Duke University Press.

Goode, Erica. 2002. "Thousands in Manhattan Needed Therapy After Attack, Study Finds." *New York Times,* March 28.

Harrison County Emergency Communications Commission. 2002. *Combined Emergency Communications and Operations Project.* Gulfport, MS. February.

Hedtke, Lorraine. 2002. *A Narrative Approach to Death, Dying and Grief.* Baton Rouge, LA: A Clinical Training Workshop.

Henderson, Captain Rodney S. 1980. "The WC-130 Meteorological System and Its Utilization in Operational Weather Reconnaissance." Air Weather Service (MAC), Scott Air Force Base, Illinois, Publication AWS/TR-80/002.

Hendrickson, Paul. 2005. "Mississippi Yearning." *New York Times,* January 10, Op-Ed.

Hoffman, Ross. N. 2004. "Controlling Hurricanes." *Scientific American* 291 (4): 68–75.

Holen, Are. 1990. *A Long-Term Study of the Survivors from a Disaster.* Oslo, Norway: University of Oslo.

Horton, James O., and Lois E. Horton. 2001. *"Hard Road to Freedom": The Story of African America.* New Brunswick, NJ: Rutgers University Press.

Hutton, J. 1976. "The Differential Distribution of Death in Disaster: A Test of Theoretical Propositions." *Mass Emergencies* 1:261–66.

The Impact of Hurricane Camille: A Storm Impact Symposium to Mark the 30th Anniversary. 1999. University of New Orleans, August 17–18.

Jeansonne, Glen. 1977. *Leander Perez: Boss of the Delta.* Baton Rouge: Louisiana State University Press.

Jelesnianski, C. P., J. Chen, and W. A. Shaffer. 1992. "SLOSH: Sea, Lake and Overland Surges from Hurricanes." NOAA Technical Report NWS 48.

Jenkins, Peter. 1995. *Along the Edge of America.* Nashville: Rutledge Hill Press.

Jordan, Nikki, et al. 2004. "Mental Health Impact of 9/11 Pentagon Attack: Validation of a Rapid Assessment Tool." *American Journal of Preventive Medicine* 26 (4): 284–93.

Klass, Dennis, Phyllis Silverman, and Steven Nickman, eds. 1996. *Continuing Bonds: New Understandings of Grief.* Philadelphia: Taylor and Francis.

Kreps, Gary A. 1984. "Sociological Inquiry and Disaster Research." *Annual Review of Sociology* 10:309–30.

———, ed. 1989. *Social Structure and Disaster.* Newark: University of Delaware Press.

Krug, Etienne, et al. 1998. "Suicide after Natural Disasters." *New England Journal of Medicine* 338:373–78.

Kubler-Ross, Elisabeth. 1969. *On Death and Dying.* New York: Scribner.

La Greca, Annette M., et al., eds. 2002. *Helping Children Cope with Disasters and Terrorism.* Washington, DC: American Psychological Association.

La Violette, Paul. 2001. *Where the Blue Herons Dance: New Tales from the Mississippi Gulf Coast.* Waveland, MS: Annabelle Publishing.

Leftwich, Preston W., Jr. 1983. "The Miss/Hit Ratio—An Estimate of Reliability for Tropical Cyclone Track Predictions." NOAA Technical Memorandum NWS 20, National Hurricane Center, April.

Lystad, Mary, ed. 1988. *Mental Health Response to Mass Emergencies: Theory and Practice.* New York: Brunner/Mazel.

Marwit, Samuel, and Dennis Klass. 1996. "Grief and the Role of the Inner Representation of the Deceased." In *Continuing Bonds: New Understandings of Grief,* ed. Dennis Klass, Phyllis Silverman, and Steven Nickman. Philadelphia: Taylor and Francis.

Mary Ann Gerlach v. State of Mississippi. No. 54,694. 1985. Supreme Court of Mississippi. 466 So. 2d 75; 1985 Miss. March 20. Appeal from Circuit Court, Harrison County; Kosta N. Vlahos, Judge.

Mason, Gilbert R., with James Patterson Smith. 2000. *Beaches, Blood, and Ballots: A Black Doctor's Civil Rights Struggle.* Jackson: University Press of Mississippi.

Menard, Donald. 1999. *Hurricanes of the Past: The Untold Story of Hurricane Audrey.* 2d ed. Cameron, LA: Menard.

Moody, Anne. 1968. *Coming of Age in Mississippi.* New York: Dell.

Morgan, Arthur E. 1971. *Dams and Other Disasters: A Century of the Army Corps of Engineers in Civil Works.* Boston: Porter Sargent.

Morgan, Elizabeth. 1998. *One August Day.* Midlothian, VA: Van Neste Books.

"Natural Features Caused by a Catastrophic Storm in Nelson and Amherst Counties, Virginia." 1969. *Virginia Minerals.* Special Issue. Charlottesville, VA: Department of Conservation and Economic Development, Division of Mineral Resources. October.

Parker, Joseph B., ed. 2001. *Politics in Mississippi.* 2d ed. Salem, WI: Sheffield Publishing.

Peacock, Walter G., Betty H. Morrow, and Hugh Gladwin, eds. 1997. *Hurri-*

cane Andrew: Ethnicity, Gender, and the Sociology of Disasters. New York: Routledge.

Perry, Ronald. W. 1979. "Evacuation Decision-Making in Natural Disasters." *Mass Emergencies* 4:25–38.

———. 1982. *The Social Psychology of Civil Defense.* Lexington, MA: D.C. Heath.

Perry, R. W., and M. K. Lindell. 1991. "The Effects of Ethnicity on Evacuation Decision-Making." *International Journal of Mass Emergencies and Disasters* 9 (1): 47–68.

Perry, Ronald W., and Alvin H. Mushkatel. 1986. *Minority Citizens in Disasters.* Athens: University of Georgia Press.

Pielke, R. A., Jr. 1999. "Hurricane Forecasting." *Science* 284:1123.

Pielke, R. A., Jr., and C. W. Landsea. 1998. "Normalized Hurricane Damages in the United States 1925–1995." *Bulletin of the American Meteorological Society* 13:621–31.

Pielke, R. A., Jr., and R. A. Pielke Sr. 1997. *Hurricanes: Their Nature and Impacts on Society.* New York: John Wiley.

Pollard, Oliver A., Jr. 1997. *Under the Blue Ledge: Nelson County, Virginia.* Richmond: Dietz Press.

Powell, Mark D., and S. D. Aberson. 2001. "Accuracy of United States Tropical Cyclone Landfall Forecasts in the Atlantic Basin (1976–2000)." *Bulletin of the American Meteorological Society* 82 (12): 2749–67.

Quarantelli, E. L. 1994. "Looting and Antisocial Behavior in Disasters." Preliminary Paper No. 205. Newark: University of Delaware, Disaster Research Center.

Rather, Dan. 1963. "The Priest and the Politician." *CBS Special Report,* September 18.

Rohlfs, A. J. 1969. "Shipping and Hurricane Camille." *Mariner's Weather Log* 13 (6): 245–51.

Rubin, Richard. 2002. *Confederacy of Silence.* New York: Atria.

Scharff, Robert G. 1999. *Louisiana's Loss, Mississippi's Gain: A History of Hancock County, Mississippi.* Lawrenceville, VA: Brunswick Publishing.

Sheets, Bob, and Jack Williams. 2001. *Hurricane Watch: Forecasting the Deadliest Storms on Earth.* New York: Vintage Books.

Shrake, Edwin. 1970. "The Lady Was a Killer." *Sports Illustrated,* March, 60–68.

Simpson, Paige S., and Jerry H. Simpson Jr. 1970. *Torn Land.* Lynchburg, VA: J. P. Bell.

Simpson, Robert H. 1970. "A Reassessment of the Hurricane Prediction Problem." Weather Bureau, National Hurricane Center, ESSA Technical Memorandum WBTM SR-50, February.

———. 1971a. "The Decision Process in Hurricane Forecasting." NOAA Technical Memorandum NWS SR-53, January.

———. 1971b. "Atlantic Hurricane Frequencies Along the U.S. Coastline." NOAA Technical Memorandum NSW TM SR-58, June.

———. 1973. "A Decision Procedure for Application in Predicting the Landfall of Hurricanes." NOAA Technical Memorandum NWS SR-71, August.

Simpson, Robert H., and D. C. Gaby. 1970. "The Satellite Applications Section of the National Hurricane Center." ESSA Technical Memorandum WBTM SR-51, September.

Spencer, Elizabeth. 1991. *On the Gulf.* Jackson: University Press of Mississippi.

Spencer, Otha C. 1996. *Flying the Weather: The Story of Air Weather Reconnaissance.* Campbell, TX: Country Studio.

Steinberg, Ted. 2000. *Acts of God: The Unnatural History of Natural Disaster in America.* New York: Oxford University Press.

Stratton, Ruth M. 1989. *Disaster Relief: The Politics of Intergovernmental Relations.* Lanham, MD: University Press of America.

Strobe, Margaret, et al. 1996. "Broken Hearts or Broken Bonds." In *Continuing Bonds: New Understandings of Grief,* ed. Dennis Klass, Phyllis Silverman, and Steven Nickman. Philadelphia: Taylor and Francis.

Sugg, Arnold L. 1969. "A Mean Storm Surge Profile." U.S. Department of Commerce, ESSA Technical Memorandum WBTM SR-49.

Sullivan, Charles L. 1986. *Hurricanes of the Mississippi Gulf Coast.* Biloxi: Gulf Publishing.

Tannehill, Ivan Ray. 1957. *The Hurricane Hunters.* New York: Dodd, Mead.

Taylor, Joe Gray. 1984. *Louisiana: A History.* New York: W. W. Norton.

Tidwell, Mike. 2003. *Bayou Farewell: The Rich Life and Tragic Death of Louisiana's Cajun Coast.* New York: Pantheon.

USACE (United States Army Corps of Engineers). 1970a. *Report on Hurricane Camille, 14–22 August, 1969.* Mobile District.

———. 1970b. *Report on Hurricane Camille, 14–22 August, 1969.* New Orleans District.

———. 1970c. *Report on Hurricane Camille, 14–22 August, 1969.* Virginia District.

United States Senate. 1970a. *Federal Response to Hurricane Camille: Part 1.* Hearings before the Special Subcommittee on Disaster Relief of the Committee on Public Works, United States Senate, Ninety-first Congress, 2d Sess. Biloxi, Mississippi, January 7. Washington, DC: U.S. Government Printing Office.

———. 1970b. *Federal Response to Hurricane Camille: Part 2.* Hearings before the Special Subcommittee on Disaster Relief of the Committee on Public Works, United States Senate, Ninety-first Congress, 2d Sess. Biloxi, Mississippi, January 8 and 9. Washington, DC: U.S. Government Printing Office.

———. 1970c. *Federal Response to Hurricane Camille: Part 3.* Hearings before the Special Subcommittee on Disaster Relief of the Committee on Public Works, United States Senate, Ninety-first Congress, 2d Sess. Roanoke, Virginia, February 2 and 3. Washington, DC: U.S. Government Printing Office.

———. 1970d. *Federal Response to Hurricane Camille: Part 4.* Hearings before the Special Subcommittee on Disaster Relief of the Committee on Public Works, United States Senate, Ninety-first Congress, 2d Sess. S. 3619: A Bill to Create, within the Office of the President, an Office of Disaster Assistance, and S. 3745: A Bill to Amend Existing Federal Disaster Assistance Legislation, and for Other Purposes, April 27, 28, and 29. Washington, DC: U.S. Government Printing Office.

———. 1970e. *Federal Response to Hurricane Camille: Part 5.* Hearings before the Special Subcommittee on Disaster Relief of the Committee on Public

Works, United States Senate, Ninety-first Congress, 2d Sess. S. 3619: A Bill
to Create, within the Office of the President, an Office of Disaster Assistance,
and S. 3745: A Bill to Amend Existing Federal Disaster Assistance Legislation,
and for Other Purposes, May 21 and 22. Washington, DC: U.S. Government
Printing Office.

Waldrop, M. Mitchell. 1992. *Complexity: The Emerging Science at the Edge of
Order and Chaos.* New York: Touchstone.

Weaver, J. D., et al. 2000. "The American Red Cross Disaster Mental Health
Services: Development of a Cooperative, Single Function, Multidisciplinary
Service Model." *Journal of Behavioral Health Service Research* 27 (3): 314–20.

Weinstein, N. 1989. "Effects of Personal Experience on Self-Protective Behavior." *Psychological Bulletin* 105:31–50.

Wilkinson, K. P., and P. J. Ross. 1970. *Citizens' Responses to Warnings of Hurricane Camille.* Report 35, Social Science Research Center, Mississippi State
University, State College.

Williams, T. H. 1969. *Huey Long.* New York: Vintage Books.

Zebrowski, Ernest, Jr. 1997. *Perils of a Restless Planet: Scientific Perspectives on
Natural Disasters.* New York: Cambridge University Press.

Zubenko, Wendy N., and Joseph A. Capozzoli, eds. 2002. *Children and Disasters: A Practical Guide to Healing and Recovery.* New York: Oxford University Press.

Newspapers

Cameron Parish Pilot. Cameron, LA (weekly), June and July 1997.

Clarion-Ledger. Jackson, MS, August 15, 2004.

Daily Herald. Biloxi, MS, August 15–September 30, 1969.

Monroe News-Star. Monroe, LA, August 3, 2003, and September 10, 2004.

Morning Paper. Ruston, LA, June 4, 2004.

Nelson County Times. August 17, 1989.

New York Times, March 20, 2004, and August 21, 2002.

Sun-Sentinel. Miami, FL, June 24, 2001.

Times-Picayune. New Orleans, LA, August 5–September 30, 1969, and
November 4–7, 2001.

Washington Post. January 16, 2005.

Interviews

LOUISIANA

Luke Petrovich, former director of public safety and Plaquemines Parish
commissioner

Leander Jurjevich, Lea Jurjevich Jr., and Charmain Jurjevich, survivors,
Plaquemines Parish

Judge Michael E. Kirby, Fourth Circuit Court, Louisiana

Marc Levitan, associate professor of civil engineering and director of Hurricane
Center, Louisiana State University

John Pine, director of Disaster Science and Management Program, Louisiana
State University

Paul Vogt Sr., Mississippi River pilot

Lucretia (alias), survivor, New Orleans

Geneva Griffith, survivor of Hurricane Audrey, Oak Grove

Steven and Florence Broussard, survivors of Hurricane Audrey, Pecan Island
Ronald Durr, survivor, Covington
Hillary Turner, retired teacher, Plaquemines Parish
Burghat Turner, resident, Plaquemines Parish
Juanita Rougeau, retired, Department of Education, Baton Rouge
Jan Shoemaker, Mississippi native, now director of service learning at Louisiana
    State University
George Tewell, Seabee during Camille, now a Louisiana resident
Barnard Schoenberger, former sheriff of Plaquemines Parish

MISSISSIPPI
H. Ben Duckworth Jr., survivor in Pass Christian, now in Jackson
Josephine Duckworth, mother of Ben Duckworth, Jackson
Bobby and Missy Goff, survivors in Pass Christian, now in Flora
Marian Martin, director of Gulfside United Methodist Assembly, Waveland
Rev. Donald Peters Sr., minister, Waveland
Charles Johnson, survivor, Bay St. Louis
Paula Isabel, survivor, Gulfport
Richard Rose, survivor, Gulfport
Gloria Payne, survivor, Bay St. Louis
Greg Durrschmidt, survivor, Keesler Air Force Base, now in Florida
George Watson, survivor and principal of Randolph School in Pass Christian,
    later assistant superintendent of schools

VIRGINIA
Bill Whitehead, former sheriff, Nelson County
Cliff Wood, former county supervisor, Nelson County
Warren Raines, survivor, Nelson County
Sharon Raines, Nelson County
Tommy and Adelaide Huffman, survivors, Nelson County
Dr. George Criswell, dentist involved in body identification, Nelson County
Phil Payne, survivor, now commonwealth attorney, Nelson County
Carolyn Albritton, survivor, Nelson County
Cecilia Epps, survivor, Nelson County
Clardine Hudson, survivor, Nelson County
Barbara Page, survivor, Nelson County
Earl Napier, survivor, Amherst
Hughes Swain, agricultural agent and coordinator of civil defense, Nelson
    County
Ed Tinsley, Virginia State Police, Appomatox

OTHER GEOGRAPHICAL AREAS
Robert H. Simpson, former director of National Hurricane Center, now in
    Washington, D.C.

Oral Histories
Recorded by R. Wayne Pyle, Mississippi Oral History Program, University of
Southern Mississippi:
    Wade Guice, civil defense director of Harrison County

Mary Ann Gerlach, survivor of Richelieu Apartments in Pass Christian, Mississippi

Paul Williams Sr., survivor who lost most of his family in Pass Christian, Mississippi

Henry Maggio, physician, Bay St. Louis, who later became a psychiatrist and testified in Mary Ann Gerlach trial

Recorded by Charlotte Capers, Mississippi Oral History Program, University of Southern Mississippi:
Olin Clark, survivor, Pass Christian, Mississippi

Recorded by Dorothy Maury, Nelson County Library archives, Virginia:
Captain Kimball Glass, chief rescue officer, state of Virginia
Paul Mays, director of public welfare, Nelson County
Sam Egleston, attorney and county coordinator, HUD, Nelson County
Russell Huffman, survivor, Nelson County

Index